T0215492

Climate Actions

Environmental and Societal Dimensions of Sustainable Development Goals

There is a global need for a structured examination of a number of environmental science projects that are producing innovative approaches to providing governments, communities, and people with outcomes from the specific themes of the United Nations Sustainable Development Goals (UN SDG) 2030 initiative. This book series takes a new and comprehensive look at interdisciplinary ways of applying environmental science, technology, and science communication to explain how environmental science serves the societal needs of specific communities around the world to achieve Sustainable Development Goals. Each of the books in this series directly explores the use of environmental science, technology, and associated approaches to environmental management in a range of biophysical, societal, and theoretical-scientific topics, specifically related to energy, science and other partnerships, food security, societal productivity and infrastructure, water resources, land resources, climate, ecology, and sustainable practices. This series provides a multi-volume resource for ecologists, engineers, natural resource planners, and professionals in other related disciplines at the local, regional, national, and global levels of decision making.

Series Editor
Ricardo D. Lopez

Climate Actions
Local Applications and Practical Solutions
Brenda Groskinsky

Rangeland Sustainability
Social, Ecological, and Economic Assessments
Kristie Maczko, John Tanaka, Aaron Harp, and Matthew Clark Reeves

Climate Actions

Local Applications and Practical Solutions

Edited by
Brenda Groskinsky

CRC Press
Taylor & Francis Group
Boca Raton London New York

CRC Press is an imprint of the
Taylor & Francis Group, an **informa** business

First edition published 2022
by CRC Press
6000 Broken Sound Parkway NW, Suite 300, Boca Raton, FL 33487-2742

and by CRC Press
2 Park Square, Milton Park, Abingdon, Oxon, OX14 4RN

Library of Congress Cataloging-in-Publication Data

Names: Groskinsky, Brenda, editor.
Title: Climate actions : local applications and practical solutions /
edited by Brenda Groskinsky.
Other titles: Climate actions (CRC Press)
Description: First edition. | Boca Raton : CRC Press, [2022] | Series:
Environmental science and the societal dimensions of global sustainable
development goals | Includes bibliographical references and index.
Identifiers: LCCN 2021047043 | ISBN 9780367478339 (hbk) | ISBN
9780367500788 (pbk) | ISBN 9781003048701 (ebk)
Subjects: LCSH: Climate change mitigation--Citizen participation. | Climate
change mitigation--Government policy--United States.
Classification: LCC TD171.75.C54 2022 | DDC 363.738/746--dc23
LC record available at https://lccn.loc.gov/2021047043

ISBN: 9780367478339 (hbk)
ISBN: 9780367500788 (pbk)
ISBN: 9781003048701 (ebk)

DOI: 10.1201/9781003048701

Typeset in Times
by Deanta Global Publishing Services Chennai, India

To my two daughters, Kacey Link and Lexi Brady.
They are the shining stars of my life and they
shine to make the world a better place.

Contents

Acknowledgments

Please allow me to acknowledge all of the people involved in this very special book, *Climate Actions: Local Applications and Practical Solutions*. It really is a "one of a kind" effort.

During my career with the U.S. EPA. I met four of the lead authors via collaborative associations with them and their organizations (Iowa State University, Bridging The Gap, The Minnesota Land Trust, and the Flathead Indian Reservation). Thank goodness, the stars were in line for the other two lead authors. I saw a presentation on the Grasslands of the Konza Prairie at the Lawrence, Kansas Public Library. The presentation was fantastic and contains material that is often overlooked related to special biological systems. So, I reached out to the presenter from Kansas State University, and he accepted. Lastly, one of my dear colleagues from the U.S. EPA provided me with a potential lead author who was studying California's urban water systems. I'm so grateful that all of the lead authors said, "Yes!"

I also want to acknowledge all of the contributors who accepted our invitation in the documentation of these very unique and special stories, during this very crazy time. Climate change is slapping us all in the face right now, but we are on a path for the future. We are working to make the planet a livable place for all of us. Hopefully, each of the chapters will inspire others to do the same.

I also want to acknowledge and thank my dear friend and colleague, Ricardo D. Lopez, who convinced me to get involved in the scientific book world and the production of much needed documentation of environmental work efforts related to environmental science and the societies that are definitely part of the process. I truly believe that if we don't document our efforts, they become lost forever. Additionally, I sincerely thank Ric for allowing me to insert the social component into the science world. Those two things can't live without each other. And I think we all are realizing that now, especially now.

Lastly, I acknowledge all of the hard work that is accomplished by the publisher's team. We could not do any of this on our own. I ask them questions on a daily basis and no one ever complains.

Congratulations to everyone involved. It was a lot of work and took up a lot of time. The final product will be worth it, not just for us, but for those who will read our book.

Series Preface

As humankind has moved fully into the 21st Century, societies have shifted much of their infrastructural and scientific foci, by necessity, to not merely "environmental awareness," but additionally to the very immediate and practical interconnections between our environs and how that environs impacts fundamental needs. The fundamental needs of humans cover the gamut, generally, from human safety and well-being, physical and other infrastructural elements, to the associated sociological components that define and shape our varied and complex communities around the world.

This book series and its various volume contributors seek to answer several practical questions that arise regarding sustainability, specifically focusing on the fundamental needs of humans, which are increasingly posing new challenges and opportunities every day to humanity. As humans ever-presently endeavor to address their fundamental needs in a changing environment, and as you read this book, you will notice a common theme, which is a longstanding desire by all of humanity to obtain a livable, comfortable, and yes, perhaps even an enjoyable life for this generation those of future generations – this fundamental desire and pursuit is generally referred to as human well-being. Humankind in its varied sociological forms around the world shares this fundamental value and desire for well-being in ways that are expressed diversely, in as varied of ways as the grains of sand among the many beaches of our planet.

From the institutional perspective on such topics as are addressed in this series, humankind has come together in a number of ways to find solutions to sustaining ourselves. One of these institutional structures that represents most of the many and varied societies of the planet is the United Nations (UN). In recent years the UN has undertaken a deliberate focus on the key elements of human well-being in their established institutionally-structured manner, by way of the UN Sustainable Development Goals (UN SDGs). Among the seventeen articulated UN SDGs are the following ten UN SDGs that focus on topics that tie directly to the environmental sciences and engineering, which is the primary focus of this series:

- Eliminating Hunger (UN SDG 2);
- Ensuring Clean Water and Sanitation (UN SDG 6);
- Developing Clean Energy (UN SDG 7);
- Stimulating Industries, Innovation, and Infrastructure (UN SDG 9);
- Supporting Sustainable Cities and Communities (UN SDG 11);
- Enabling Responsible Consumption and Production (UN SDG 12);
- Informing Climate Action (UN SDG 13);
- Protecting Life below Water (UN SDG 14);
- Fostering Life on Land (UN SDG 15); and
- Developing Partnership for Meeting Sustainable Development Goals (UN SDG 17).

This series seeks to explore each of the ten UN SDGs listed above through the lens of ongoing and future work in the environmental sciences and engineering, integrating the importance of science partnership with specific communities and practitioners around the world.

The series volume you are currently exploring, *Climate Actions: Local Applications and Practical Solutions*, is an excellent demonstration of how sound science, passionate communities and practical applied methods are being implemented at local, regional and transboundary scales to directly address UN SDG 13 (Informing Climate Action). This volume covers a vast "landscape" of science and engineering topics by seamlessly integrating human needs, cultural norms and practices and a diverse range of geographic settings. The editor and authors of this volume have taken great care to ensure that not only the demonstrations of "action" are clear and repeatable within the reader's community of practice, profession and/ or professional circumstances, they have also outlined the various pitfalls that have beset a number of scientific and societal endeavors in the past, with regard to implementation. Consequently, this volume serves as an excellent guide for the both the experienced climate scientist and those who may be struggling to find "climate action" solutions within their particular community, region or organization. Indeed, this volume affords opportunities for common focus among climate science professionals and practitioners who currently, or are planning to, work together on challenging climate science issues.

This volume on climate action seeks to solve one of the most pressing and existentially important sustainable development goals for our planet, and indeed succeeds in guiding us all through various contemporary views, including a number of traditional ways of knowing, which may prove as important as ever for maintaining and improving humankind's existential needs now, and into the future.

Ricardo D. Lopez, PhD (Series Editor)
October 2, 2021

Preface

Climate Actions: Local Applications and Practical Solutions is a part of a newly created international series of books based on the United Nations Sustainable Development Goals (UN SDGs). Each of these goals, such as "Informing Climate Action," is the premise for the themes of each book in the series. Our book's chapter authors developed six specific climate action chapters. Each chapter contains a purposeful theme that documents how they created a climate action (or a set of actions). All of the climate actions that were created by this very diverse set of authors are strikingly distinctive.

Through the use of various environmental science and related disciplines, each chapter will initially document the authors' experiences witnessing climate change and the environmental and human health problems that have been realized as a result. In specific locations in the United States and abroad, each chapter will articulate how this global issue is altering the earth's natural and built landscapes that ultimately is hindering our abilities to provide adequate human health and social well-being.

Notably, there actually is a large number of climate actions that are currently in process worldwide. Therefore, we are asking you to please draw your attention to the book chapters' individual set of very unique and socially creative climate actions. Once the climate action chapters have been read, hopefully you will be inspired to create your own specific climate actions in your neighborhoods or communities. We expect that the set of climate actions in this book will inspire you to tell others about your climate efforts supporting the quest of all of us to work toward the elimination of climate change before it becomes a global catastrophe.

Most importantly, the individual chapters will uniquely portray the novel efforts of teams of environmental research scientists, environmental and human health policy experts, nonprofit organizations and Indigenous peoples, noting that the implementation of a climate action or a set of climate actions requires a social construct. The collective set of "Climate Actions" in this book has become a means toward climate mitigation by the authors, in the specific places where they live. Lastly, each chapter's documented conclusion will articulate the resulting solutions and/or potential outcomes.

Brenda Groskinsky, August 2021

For more information about the United Nations' Sustainable Development Goals (UN SDGs), please go to: https://sustainabiledevelopment.un.org/sdgs

Editor

Brenda Groskinsky has an extremely diverse education, and now that she is retired from federal service (USGS and U.S. EPA), she is enjoying the documentation of science-based collaborations while focusing on the societal dimensions of environmental topics. As the appointed Science Policy Advisor for the U.S. EPA, Brenda implemented numerous research-based interactions with international, local, state, military and tribal organizations. As an example, Brenda provided technical assistance for the U.S. EPA's Office of International Activities in the Czech Republic and Slovakia after the fall of the Velvet Revolution (1993–1999). Brenda helped to facilitate the implementation of an innovative biological treatment process for the removal of ammonia, arsenic, iron and manganese from a small drinking water system in Iowa. Additionally, Brenda initiated the U.S. EPA support for the NetZero partnership with Ft. Riley, Kansas, and was a key player in the installation of an anaerobic membrane bioreactor at the facility. Brenda also was elected as a co-chair of the U.S. EPA National Tribal Science Council, working with Native American Tribes and Organizations. Brenda holds M.S. degrees in Applied Mathematics (Portland State University) and Computer Networking (University of Missouri-Kansas City) and an M.S. equivalent in Information Resource Management (National Defense University). Additionally, she has pursued advanced graduate work in Economics and has passed PhD comprehensive exams in Numerical Analysis and Statistics (University of Missouri-Kansas City). Brenda seeks to continue pursuing and documenting successful applied science policy research collaborations while highlighting the connections between the cultural, social and economic components of environmental topics.

Contributors

Seton Bachle
Department of Forest and Rangeland
    Stewardship
Colorado State University and Division
    of Biology
Kansas State University
Manhattan, KS

Nicholas Bancks
The Minnesota Land Trust
Saint Paul, MN

Virginia Breidenbach, PE
The Minnesota Land Trust
Duluth, MN

Anne A. Carlson
The Wilderness Society
Choteau, MT

Richard M. Cruse
Iowa State University
Ames, IA

Rene Dubay
Native Tracks Education Institute
Polson, MT

Michael Durglo, Jr.
Confederated Salish and Kootenai
    Tribes (CSKT)
Pablo, MT

Richard G. Everett
Salish Kootenai College
Pablo, MT

Fernando García-Préchac
Faculty of Agronomy
Republic University of Uruguay
Montevideo, Uruguay

Haley Golz
The Minnesota Land Trust
Duluth, MN

Brenda Groskinsky (Editor)
Lawrence, KS

Sasha Harris-Lovett
Berkeley Water Center, UC Berkeley
Berkeley, CA

Emily Heaton
College of Illinois at
    Urbana-Champaign
University of Illinois
Urbana, IL

Tony Incashola, Jr.
Confederated Salish and
    Kootenai Tribes
Climate Change Advisory Committee
Pablo, MT

Rachel Keen
Division of Biology
Kansas State University
Manhattan, KS

Kris Larson
The Minnesota Land Trust
Saint Paul, MN

Baoyuan Liu
Beijing Normal University
Beijing, China

Richard G. Luthy
Department of Civil & Environmental
    Engineering, Stanford University
Stanford, CA

Maureen I. McCarthy
Native Waters on Arid Lands
Desert Research Institute
Reno, NV

Jesse Nippert
Division of Biology
Kansas State University
Manhattan, KS

Wayne Ostlie
The Minnesota Land Trust
St. Paul, MN

Panos Panagos
European Commission
Joint Research Centre (JRC)
Ispra (VA), Italy

ShiNaasha H. Pete
Confederated Salish and
    Kootenai Tribes
Ronan, MT

Net Phillips
The Minnesota Land Trust
Saint Paul, MN

Beverly Rinke
The Minnesota Land Trust
Saint Paul, MN

Kristin Riott
Bridging The Gap
Kansas City, MO

Joshua Rosenau
Center for Tribal Research & Education
    in Ecosystem Sciences (TREES)
Salish Kootenai College
Pablo, MT

Ruurd Schoolderman
The Minnesota Land Trust
Saint Paul, MN

Séliš-Qlispé Elders Council
Confederated Salish & Kootenai Tribes
St. Ignatius, MT

Thompson Smith
Tribal History & Ethnogeography
    Projects
Séliš-Qlispé Culture Committee
Confederated Salish & Kootenai Tribes
St. Ignatius, MT

Dennis Todey
USDA-ARS Midwest Climate Hub
National Laboratory for Agriculture and
    the Environment
Ames, IA

Shirley Trahan
Séliš-Qlispé Culture Committee
St. Ignatius, MT

Chunmei Wang
Northwest University
Xi'an, Shaanxi, China

Enheng Wang
Northeast Forestry University
Harbin, Heilongjiang, China

Emily Wedel
Division of Biology
Kansas State University
Manhattan, KS

People are the cause of climate change;

our actions or inactions are the cause of climate change.

Mother earth can and will survive without us,

we cannot survive without her.

Michael Durglo Jr.
Climate Change Advisory Committee Chair
Confederated Salish and Kootenai Tribes

1 Our Local and Practical Places

Brenda Groskinsky

CONTENTS

1.1 INTRODUCTION OF THE LOCAL AND PRACTICAL

The two words, local and practical, should create an opportunity for us to think about the place(s) where we live, and that they are the places that accommodate our needs. Before now, none of us needed to think about how the places where we live are "changing" right in front of us. But we certainly have to think about it now. I live in the state of Kansas. For the last couple of weeks, it has been too smoky and too hot to go outside, for even a short amount of time.[1] And we don't even have fires burning here. The U.S. fires burning on the West Coast, as I write this, are bringing the smoke to us. I used to run three times a week. I can't do that now. The air quality has not been healthy for me anymore. Instead, I have to use my indoor elliptical for cardio exercise, noting my air purifier is operating right next to me. I'm now wondering how do we get back to our normal, local and practical lives? For most of us, the thought of changing how we live because of the impacts of global warming and climate change seems very daunting.

1.2 THE ADVANCEMENT OF GLOBAL WARMING

Humans have caused major climate changes to happen already, and we have set in motion more changes still. However, if we stopped emitting greenhouse gases today, the upward rise in global temperatures would begin to flatten within a few

DOI: 10.1201/9781003048701-1

years. Temperatures would then plateau, but remain well-elevated for many, many centuries to come. There is a time lag between what we do and when we feel it, but that lag time is less than a decade. While the effects of human activities to date on Earth's climate are irreversible on the timescale of human lifetimes, every little bit of avoided future temperature increase represents less warming that would otherwise persist for essentially forever. The climate benefits of greenhouse gas emissions reductions occur on the same timescale as the political decisions that lead to those reductions (Shaftel, 2021).

1.3   NEW BOOK SERIES BASED ON THE UNITED NATIONS SUSTAINABLE DEVELOPMENT GOALS (SDGS)

CRC Press has embarked in the development of a new book series, *Environmental and Societal Dimensions of Sustainable Development Goals*, based on the United Nations' Sustainable Development Goals (UN SDGs). Ten of the goals match up with ten books in this series of books. The book you are reading now, *Climate Actions: Local Applications and Practical Solutions*, is based on the United Nations' *SDG of Informing Climate Action*.

1.4   WHAT IS A CLIMATE ACTION?

Actually, there truly is an unlimited number of climate actions, and each action that is implemented gets us closer to the positive modification of climate change. The beauty of implementing a climate action is that even a small action can make a positive difference in all of our lives. For example, taking the time to insulate your basement that creates enhanced energy efficiency for your home and thus, providing lower energy bills and increased warmth for your family. Another climate action example is the implementation of wind turbines as a way to generate electricity instead of using fossil fuel–based resources that create carbon dioxide emissions.

> A single rotation of the blades generates the electricity for one household's daily use. … In the United States, the wind energy potential of just three states – Kansas, North Dakota, and Texas – would be sufficient to meet electricity demand from coast to coast.
> **(Hawken, 2017)**

1.5   HOW ARE CLIMATE ACTIONS CONNECTED TO CLIMATE CHANGE?

The word *climate* "is a statistical description of the weather over a period of time, usually a few decades" (Dessler, 2016). Therefore, "'climate change' refers to increasing changes in the measures of climate over a long period of time – including precipitation, temperature, and wind patterns" (USGS, 2021). Any activity that works to combat climate change while reducing global warming is a climate action.

1.6   THE UTILITY OF DOCUMENTING UNIQUE CLIMATE ACTION STORIES

The distinct climate action stories that are told in this book, by authors from all over the world, clearly have a scientific component. However, it is the people on this earth who will ultimately have to make a difference in our struggles to defeat climate change. The creation and implementation of practical climate actions can work to save our local spaces and our ways of life.[2]

The authors' individual stories and documentation of peoples' lives in this book will connect you to the local places where they live, and the design and implementation of their specific practical climate action solutions, all happening in the shadow of climate change.

1.7   A SUMMARY OF THE CLIMATE ACTION STORIES

Six chapters of this book reflect extremely carefully designed climate action solutions created by a large set of authors, who are anxious to tell their stories on how they are implementing climate actions all over the world, and in very unique communities.

Vast teams of worldwide environmental research scientists, environmental and human health policy experts, nonprofit organizations and Indigenous people who have sacred knowledge have witnessed and investigated a wide variety of climate change challenges. Their challenges and their science and social solutions are documented in this book. The final result is the development and documentation of a unique collective set of "Climate Actions" in the specific places where they live.

There are a couple of climate action efforts that specifically focus on agricultural (Chapter 2) and native plant communities, such as worldwide grasslands (Chapter 4). Both describe the necessity for methodical oversight as a means to overcome alteration caused by not only climate-related issues, but also crop and grassland management processes that allow soil degradation (Chapter 2) and tree encroachment (Chapter 4).

It is not a new information that agriculture and special ecosystems require a stable climate. Because of the current climate-related issues, we are now forced to create manmade processes that will limit the degradation of soil and water resources. Chapter 2, which involves a worldwide team from China, Uruguay, Italy and the United States, has documented three separate approaches that include an incentive-based strategy, government regulatory practices, and government and regulatory land management plans.

Grasslands can be found on every continent except Antarctica. They are unique ecological systems that contribute to a number of ecosystem processes that are extremely important, such as carbon storage and regulation of the water cycle. Chapter 4 outlines a climate action plan that focuses on grassland ecosystems at multiple levels, including individual, community and global scales. I've visited Kansas's Konza Grassland Prairie, more times than I can count. I have hiked the prairie trails, always noting a new species of lizard or flower that I've not seen before. Chapter 4 offers multiple potential solutions that can enhance the viability and persistence of grasslands. They are an extremely unique and a viable necessity in the survival of our planet.

Chapter 3 articulates that many low-income people, who are living in older housing stock, are typically forced to endure the cold in the winter from the lack of heat, and then in the summer months they suffer from heat because they do not have access to air-conditioning. Perhaps a potentially obvious solution is the documentation of why so many people procrastinate implementation of energy efficiencies in buildings and homes, when they could be creating mechanisms that can resolve the social injustices. Kristin Riott, Executive Director of Bridging the Gap, a nonprofit organization in Kansas City, MO, is working to create and understand the psychological nuances of energy efficiencies, i.e., the procrastination of insulating basements. There are multiple weather-related deaths in Kansas City every year. The implementation of energy efficiencies in low-income neighborhoods is a necessary climate action.

Chapter 5 authors note that many coastal areas in California relied on snow melt from the Sierra Nevada mountains for their drinking water. However, urban water systems are now being challenged noting that the freshwater systems just can't keep up with the demand. The key question is: what do we do when all of the clean drinking water sources have dried up due to climate-related events and issues?

The authors document that the potential for stormwater capture and reuse is a current and critical climate action. They also summarize that we need to develop methods to capture stormwater as well as the creation of methods to facilitate water reuse. It will certainly require a cadre of diverse, knowledgeable and committed stakeholders to create and implement the water systems of the future. One unique opportunity described in Chapter 5 documents how the American company, Facebook, has created an onsite water reuse system for toilets. Clearly, innovative ideas and technologies are in play. The authors significantly note that if California's urban water systems are to survive, they will need to break away from importing water from the Sierra Nevada.

Chapter 6 points out that the state of Minnesota has been described as the second warmest state in the United States. How does that make any sense at all, unless climate change is involved? The nonprofit, Minnesota Land Trust, has been working to facilitate the natural adaptation of local systems, i.e., ecosystem services such as flood protection and clean water. Notably, 76% of Minnesota's landscape is privately owned. Chapter 6 describes the protection strategies of a collective group of landowners and conservation partners and points out that "these individual actions and web of relationships support the long game of landscape level conservation and the relentless incrementalism that is required to improve climate resiliency across the landscape."

Chapter 7 documents our last climate action example. The Confederated Salish-Kootenai Tribal Nation, located in Montana, are using science-based and spiritual knowledge to save their sacred Whitebark Pine Forest from climate modification. They are gathering seedlings and replanting the new trees in areas where the trees can not only survive but thrive. Their story describes implementing a unique climate action in order to save and restore their forest and their ways that have been their lives for thousands of years.

The questions and topics listed in this chapter are overwhelming and inspiring at the same time. Maybe there are things that we could implement to get us back to the practical, noting that the practical behavior implies that our efforts will be local as well as efficient.

NOTES

1. https://weather.com/science/environment/news/2021-08-13-hottest-month-on-earth-july (checked August 14, 2021).
2. https://www.nationalgeographic.com/environment/article/some-irreversible-changes-to-the-climate-can-still-be-headed-off-report-says (checked August 9, 2021).

LITERATURE CITED

Dessler, A. (2016). *Introduction to modern climate change*. New York: Cambridge University Press.

Hawken, P. (Ed.). (2017). *Drawdown: The most comprehensive plan ever proposed to reverse global warming*. New York: Penguin Books.

Shaftel, H. (2021). Global climate change. NASA's jet propulsion laboratory. California Institute of Technology. https://www.caltech.edu/ Checked August 9, 2021.

USGS. (2021). *What is the difference between global warming and climate change?* Retrieved at https://www.usgs.gov/faqs/what-difference-between-global-warming-and-climate-change-1 (accessed August 23, 2021).

2 Agricultural Adaptation to Climate Change
Limiting Degradation of Soil and Water Resources

*Richard M. Cruse, Enheng Wang, Chunmei Wang,
Fernando García-Préchac, Panos Panagos,
Baoyuan Liu, Emily Heaton, and Dennis Todey*

CONTENTS

DOI: 10.1201/9781003048701-2

2.1 INTRODUCTION

The most agriculturally productive rain-fed areas exist in regions of the world having fertile soils and climate conditions that support relatively high and stable crop productivity (Fischer et al. 2002). Further, these areas exist because unique combinations of soil-forming factors and the climate conditions favoring this development have remained stable over millenniums. Inherent in both the soil development process and current crop production is the interaction of climate, soil and vegetation; the climate was favorable for plant biomass production during the soil development process and remains favorable for crop production now. Prior to human influence, the soil and plant community system that evolved was both resilient and self-sustaining – climate and soil-favored nutrient and water uptake through plant growth; plants converted atmospheric carbon into organic matter and furnished it to soil biota that functioned to recycle nutrients in the biosphere, and the plant canopy covered the soil surface protecting it from stresses associated with periodic extreme weather conditions or events.

Current agriculture methods have altered this resilient system by replacing native plant communities with food, feed and biofuel crops. Crop management practices have altered the soil and soil surface cover reducing soil protection from rain, wind and heat. These practices have also changed the amount and distribution of biomass (aboveground vs. belowground; perennial vs. seasonal), which impacts soil organic matter dynamics, nutrient cycling and soil biology. Disrupting such an inherently resilient system carries a risk of degrading system components, especially soil (Lal, R. 2012). This is particularly concerning because most areas recognized as favorable for dryland production by Fischer et al. (2002) are also vulnerable to accelerated soil erosion (Borrelli et al. 2017) and resulting land degradation.

A stable or stationary climate has been an important component contributing to high and reliable plant production potential in these areas. While modern agriculture has played an important role in disrupting resilient biomass producing systems, the changing climate brings another existential threat to long-term productivity (IPCC 2019). Despite the rather ominous evidence projecting continued and even accelerated soil degradation rates, the most important factor impacting soil sustainability, especially for soil erosion, remains land management choices (Olsson et al. 2019). How we choose to manage our soil resources will determine the long-term productivity of our global soil resource base. This chapter will identify agricultural practices and approaches being used to adapt to a changing climate for five different agriculturally productive areas of the world – the Northeast Mollisol belt in China, the Loess Plateau in China, Europe, Uruguay and Central United States.

2.2 CHANGING CLIMATE AND SOIL DEGRADATION

The Earth and its atmosphere are warming and extremes in precipitation and drought are more frequent and intense; the scientific confidence that these trends will continue is high (IPCC 2019; USGCRP 2018). Evidence also indicates seasonal rainfall shifts are occurring, at least in selected areas; for example, in the Central United

States, increased spring rainfall and reduced rainfall in late summer and autumn are being recorded, and models suggested this is likely to continue (Feng et al. 2016). These shifts interact with timing of agricultural activities affecting soil degradation, as will be explained later. The combination of spatial and temporal climate shifts will likely add stress to soil and water resources beyond that historically experienced.

Climate components, especially precipitation and temperature, have major impacts on soils and the capacity of soil to produce agricultural crops. Most productive soils have a favorable surface layer, the A-horizon, with elevated soil organic matter and nutrient contents compared to deeper horizons. The A-horizon developed favorable properties largely because of elevated biological activity in this layer – maximum plant rooting density and associated macro- and micro-fauna that rely on root materials deposited in the soil for their life processes. The A-horizon characteristics, especially depth (see Figure 2.1), have been repeatedly used as an independent variable when characterizing soil loss (A-horizon thinning) impacts on crop yields (den Biggelaar et al. 2003).

More intense precipitation, longer growing seasons and elevated temperatures threaten the nature of the A-horizon. Longer growing seasons (longer duration of warmer temperatures) and higher temperatures lead to warmer soils; organic matter oxidation rates and thus soil organic matter loss have a direct positive relationship with temperature (Amundson et al. 2015). The soil organic matter balance, additions vs. losses, could remain positive *if* our changing climate resulted in appreciably greater organic matter deposition belowground than losses associated with temperature rise. This might occur if warmer conditions induced farmers to adopt crops with greater root biomass or if plant genetics were modified such that the current

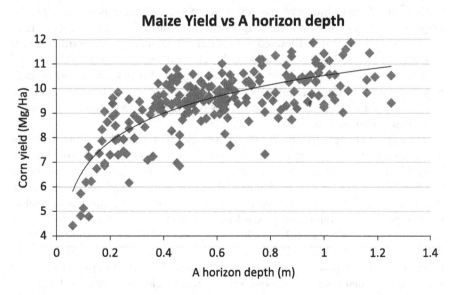

FIGURE 2.1 Relation between maize yield and A-horizon depth in Central United States across four years with growing season precipitation below normal.

crops increased root biomass, for example. However, current evidence indicates the entrenched markets and infrastructure for current commodity crops will likely preclude major changes in cropping systems. This is especially true for the United States. However, efforts are underway to modify current cropping systems such that they are more soil organic matter friendly.

Changing precipitation patterns create a different type of challenge for the productive surface layer. Elevated rainfall amounts and rainfall energy increase soil erosion potential (IPCC 2019), resulting in increased translocation of surface soil particles and materials, for example, organic matter, attached to them. Movement of soil particles downslope with subsequent deposition where the velocity of overland flowing water slows results in soil removal and A-horizon thinning of more steeply sloping areas with soil accumulation downslope. The combination of tillage-induced downslope soil movement and water erosion has resulted in approximately 35% of the farmed U.S. Corn Belt devoid of topsoil (Thaler et al. 2021). This has occurred during a climate period that has been relatively stable and favorable for farming activities (Takle and Gutowski 2020). The challenges imposed by elevated rainfall energies in the world's food-producing areas must be met with management practices that limit the negative impact of heavier rainstorms while maintaining crop productivity.

Reduction in yields associated with thinning A-horizon has been documented extensively (den Biggelaar et al. 2003). Of arguably greater importance is the impact of thinning topsoil on crop productivity during climate stress conditions. Water infiltration and storage in the soil profile are dependent on favorable soil physical conditions, especially at and near the soil surface. Removing the most favorable soil materials from the soil surface layer will reduce infiltration rates, leading to more surface runoff, and ultimately less soil water storage. Climate projections are for longer dry periods between increasingly heavy and intense rainfall events or rainfall periods (IPCC 2019). Elevated temperatures will increase evapotranspiration demand and longer times between rainfall events will increase the overall need for soils to store more water for crop use. While improved management and genetics may help increase water use efficiency, it remains a biophysical law that biological productivity for a given plant or crop species is directly related to quantity of water transpired (Ben-Gal et al. 2003).

The impact of degraded soil combined with elevated water demand required for higher crop yields poses challenges for maintaining or increasing existing productivity. Figure 2.1 illustrates the corn yield response to thinning topsoil measured on Mollisols in Central Iowa, United States. Maize yield and corresponding A-horizon depth were recorded across four years in which growing season rainfall was below normal (personal communication Tom Kaspar, USDA-ARS, 4/19/2021). As horizon depths declined to less than 0.2 m, yields rapidly declined. Because a substantial area within the U.S. Corn (Maize) Belt is now missing A-horizon (Thaler et al. 2021), the reduced production potential associated with increased heat and drought stress for degraded soil conditions appears ominous. Specific yield response to soil degradation is crop specific; however, negative production impacts of degraded soil is to be expected for virtually all crops produced in rain-fed regions, especially under climate stress conditions.

Seasonal change in precipitation and its interaction with farming operations impacts soil erosion potential in more subtle ways. In the Central United States, increasing spring precipitation associated with a changing climate (Feng et al. 2016) brings with it potential delays in pre-plant tillage and planting of row crops. While our biggest concern is typically delayed planting impacts on crop yield, these delays can also impact soil erosion rates. Figures 2.2 and 2.3 illustrate the impact of delayed tillage and planting on soil erosion, estimated using the Daily Erosion Project (Gelder et al. 2018) and observed Iowa rainfall. Spring rainfall becomes more erosive in the Central United States during April to June. The most vulnerable soil erosion condition for most row crops occurs between the time of pre-plant tillage (that reduces surface residue cover) and crop canopy closure. Delaying this vulnerable condition until later in the spring aligns with more erosive late spring rainfall.

Tillage/Planting Date	Soil Erosion (Mg/Ha)
10-Apr	4.0
20-Apr	4.3
30-Apr	4.5
10-May	4.9
20-May	5.4
30-May	5.6

FIGURE 2.2 Annual sheet and rill erosion estimates associated with progressively later springtime tillage and planning dates in Iowa, USA.

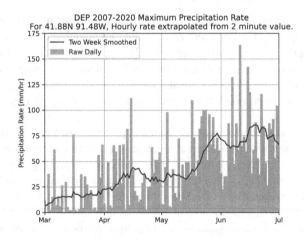

FIGURE 2.3 Maximum daily observed precipitation rate occurring from March 1 to July 1 for 41°88′ N and 91°48′ W (East–Central Iowa) across the time period 2007–2020.

2.3 US CLIMATE CHANGE ADAPTATION
AND MITIGATION STRATEGIES

Adaptation strategies are based on concepts proven to reduce soil degradation or improve soil health, which is especially important during harsh climatic conditions. Practices that provide continuous surface cover, increased plant root growth and/or root density, diversified crop rotations and limited or no tillage are the primary in-field tools for directly improving soil health and indirectly adapting to climate change in the Central United States. Of these approaches, limiting or eliminating tillage has arguably been the most successful in this era of voluntary compliance and no or limited government regulation (explained later) because tillage reduction has been cost effective, supported by industry and responsible for improved yields and better economics for many farmers.

A wide variety of state and Federal government programs encourage practices that support climate change mitigation or adaptation. Three examples of Federal government–supported programs are given below. In addition, a range of farmer-led groups and organizations, for example, Practical Farmers of Iowa demonstrate and promote practices to others that are economically favorable, that are resilient to weather anomalies and that favor soil health maintenance or improvement. A variety of nonprofit groups such as The Nature Conservancy work with researchers and farmers to increase favorable practice adoption and develop new management approaches favoring soil health. Many opportunities exist for interested farmers to engage with experts and/or obtain government financial support for adoption of practices that are perceived to adapt to and mitigate against climate change.

2.3.1 GOVERNMENT APPROACHES

The U.S. Department of Agriculture (USDA) Climate Hubs have often endorsed a concept developed by the Food and Agriculture Organization of the United Nations (Food and Agriculture Organization of the United Nations 2010) termed Climate Smart Agriculture. While there are ten building blocks associated with Climate Smart Agriculture, today's U.S. Climate Smart Agriculture has three primary goals (Buda 2021): increased productivity, enhanced resilience and reduced greenhouse gas emissions.

The USDA has shaped various agricultural programs to induce farmers to conserve soil and water with in-field practices, and as a result these programs are aligned with Climate Smart Agriculture principles and goals. Encouraged practices have typically existed for years – they are not new. Recognition that we have been operating for decades in a relatively favorable climate for agriculture (Takle and Gutowski 2020) and that with this favorable climate substantial soil degradation has occurred is adding a new level of seriousness to agriculture resilience discussions considering projected challenging climate change trends. Concerns over increased rainfall intensity and longer periods of drought and extreme heat have prompted multiple programs with increased focus on soil health.

The USDA's approach to changing or influencing farmer decisions has been and continues to be volunteer-based with little effort to modify agricultural practices through regulatory means. Volunteer in this sense means farmers choose to adopt a practice with few or no government mandates imposed regarding practices to be used. In many situations, a government agency encourages farmers to change practices through offering a government subsidy payment and/or technical assistance from sources such as county-based Natural Resource Conservation Offices (NRCS). University Extension service and farm cooperative agronomists likewise play critical roles in educating farmers regarding alternative practices. This is a sharp contrast to approaches used in some countries such as Uruguay (explained later in this chapter) that rely heavily on regulatory approaches to ensure practices are used that favor soil and water resources and climate change adaptation and mitigation.

2.3.1.1 Selected U.S. Government Program Examples

The NRCS administers through its county offices multiple conservation programs of potential interest to a wide range of producers. One of the more popular programs is the Environmental Quality Incentives Program (EQIP) (USDA, Natural Resources Conservation Service 2021a). Producers are offered financial support and technical assistance to adopt practices new to their farming operations, practices such as cover crops, extended crop rotations, forest stand improvement, prescribed grazing and new irrigation. The farmer and agency sign a binding contract regarding practice implementation and support payments for specific conservation practice implementation on targeted farmland area(s).

The Conservation Stewardship Program (CSP) administered through the local NRCS office supports farmers financially and with technical assistance for improving existing conservation efforts on their farms' working lands, that is, on land being used for crop production, grazing or production of forests. CSP is the largest agricultural conservation program in the United States (USDA Natural Resources Conservation Service 2021b).

The Conservation Reserve Program is administered by the Federal Farm Service Agency (FSA) and is designed to remove farmland sensitive to soil erosion from crop production while enhancing environmental benefits and mitigating climate change through carbon sequestration. Farmers sign a 10- or 15-year contract with the agency and for removing erosion-sensitive lands from production and growing environmentally favorable plant species, they receive an annual land rental payment from the Federal government (USDA Farm Service Agency 2021).

2.3.2 Do These Approaches Work?

The NRCS periodically estimates sheet and rill soil erosion rates for each state in the United States using Revised Universal Soil Loss Equation technology. This allows an evaluation of soil erosion rates through time as influenced primarily by crop and soil management practices. Nationally, soil erosion rate estimates decreased from 1982 to 1997 after which soil erosion rates have remained stable (U.S. Department of Agriculture 2020). The major corn-producing states in the Central United States have seen similar estimated soil erosion rate declines from 1982 until 1997; however, erosion rate increased after the 1997–2002 period (U.S. Department of Agriculture 2020), which coincides with the time that important soil conservation programs were implemented. Continued soil erosion rates greater than soil renewal rates and acceleration of these rates bring into question the capacity of existing volunteer programs to foster adoption of practices necessary to maintain resilient soil resources, resources that will be necessary in a more hostile climate.

2.3.3 CARBON MARKETS

The capacity of soils to sequester carbon is well known (Lal 2012). However, means to estimate the effect of cropping systems or selected management practices on quantity of carbon sequestered has been elusive. Nonetheless, the concept of carbon markets based on carbon offsets is gaining popularity. Farmers who use practices that, based on peer-reviewed science, are effective at sequestering carbon receive payment for an estimated mass of carbon sequestered by a given practice over a given time. Carbon markets are designed to link industries that exceed established minimum levels of carbon release to farmers who are "capturing" the industries carbon release excess. Payments made to farmers for carbon capture originate from industry and are managed through a "carbon exchange" or business that manages such exchange activities.

2.4 URUGUAY: INTRODUCTION

The Uruguayan continental area is 169,000 km^2 located between 30–35° S and 53–58° W. Annual average precipitation is 1,100 mm ($\pm$ 200 mm); mean annual temperature is 24°C in summer and 12°C in winter. No frozen soils nor snow cover exists. The country belongs to the Río de la Plata Grasslands Physiographic Unit (Paruelo et al. 2001) that occupies 65% of the territory. Topography is gently rolling; dominant slopes are 3–6%, with some flat plains and some areas with more than 8% slope; mean altitude is 140 m above sea level. In Soil Taxonomy, the most important soils are Mollisols and Vertisols, but there are significant areas of Alfisols, Ultisols, Inceptisols, Entisols and Histosols.

Durán (1998) estimated the Uruguayan soils organic carbon (SOC) content using the national general soil map (1:1 million scale, 99 mapping units). Values came from sampling and analyzing 200 profiles, mostly from undisturbed soils. Therefore, his information represents the country's potential SOC content assuming other country soils could match the SOC of the sampled profiles. Durán's results indicate that to a 1 m depth, the country's soils can hold 2.3 Pg of SOC, with a mean value of 13.4 $kg \cdot m^3$. This is 17% above the world average of 11.5 $kg \cdot m^3$ (Eswaran et al. 1993 and 1994, cited by Durán 1998). The area occupied by the mapping units dominated by Mollisols and Vertisols, representing 30.6% of the country, have SOC content from 15 to 20 $kg \cdot m^3$ and more. Also, 40–45% of these soil's SOC is in the upper 20 cm of the profile.

Until the mid-20th century, agriculture was dominated by continuous cropping (CC) with conventional tillage; wheat was the main crop managed with several tillage operations leaving low amounts of surface crop residues. This generated unacceptably high soil erosion rates, having significant negative effects on approximately 30% of the country's surface by the mid-1960s, impacting the most productive soils in the country (Cayssials et al. 1978, cited by Durán and García Préchac 2007). A complementary study (Sganga et al. 2005, cited by Durán and García Préchac 2007) identified the same proportion of the territory had been negatively affected (30.1%), separating the degradation into the following categories: slight 18.3%, moderate 9.9%, severe 1.3% and very severe 0.6%.

By the mid-1960s, there was a general adoption of crop–pasture rotations (CPR), cropping three to four years followed by another three to four years of seeded grass and legume pastures for direct grazing. This change, even with conventional tillage, resulted in an important erosion rate reduction and SOC content recovery during the pasture CPRs phase (García Préchac et al. 2004). During the 1990s, mostly no-till (NT) and some reduced till (RT) replaced conventional tillage; this, together with the CPRs, improved soil conservation further (García Préchac et al. 2004).

New cropping intensification began early in the 21st century, resulting in shorter pasture duration in the CPRs or even its elimination, resulting in a much greater CC area. This dramatic change was due to the relatively low price of Uruguayan land compared to other land in the region, excessive taxing to grain exports in Argentina and the great international increase of soybean price. Large-scale Argentina agricultural enterprises came to Uruguay, generating structural changes in land size, tenure and operational management (Arbeletche et al. 2010). Soybean became the new leading crop, increasing its area from almost nothing, around 10 kilo hectares (kha) in 1999 to 14,000 km^2 in 2014.

This created major challenges for Uruguay soils as little experimental data existed addressing soybean effects on soil degradation; soybean was not previously grown in Uruguay. Thus, models (USLE/RUSLE to estimate erosion and CENTURY to estimate SOC) were used to estimate management effects on SOC. Their predictions indicated that CC with NT soybean monoculture is not sustainable due to its erosion rate and loss of SOC. Including winter cover crops or double annual cropping of soybeans and wheat could reduce erosion close to tolerance (typically 7 $Mg \cdot ha^1 \cdot yr^1$), but could not totally balance the loss of SOC. Sustainability could be achieved with CPR-NT systems, producing erosion rates similar to the ones under natural grasses and maintaining or modestly increasing SOC content. The first modeled prediction was confirmed empirically with the mean of about 4,000 soil samples analyzed yearly in the soil test lab of INIA (Exp. Station La Estanzuela) between 2002 and 2014 (Beretta et al. 2019); the majority of these samples came from continuously cropped fields with soybean. An approximately 20% relative drop in SOC, along with lower content of exchangeable bases and lower soil pH, was observed. When SOC falls, the cation-exchange capacity falls, which causes the loss of bases and, consequently, the acidification of the soils.

The political reaction to the new agricultural reality, considering that conserving soil is required to maintain soil quality and productivity, was to update and effectively apply the country's soil conservation legislation. The objective of this section is to describe this policy and its consequences in terms of soil and SOC conservation. The conservation or even recovery (sequestration) of SOC is the main contribution that the country can make to mitigate climate change, in addition to reducing fossil fuel consumption. In terms of adaptation, evidence indicates that SOC conservation is reached with more diverse production systems that are more resilient to climate change and world market variations.

This section is mainly composed of presentations in the Global Symposium on Soil Organic Carbon, March 2017, Food and Agriculture Organization, Rome, Italy (García Préchac et al. 2017), Global Symposium on Soil Erosion, May 2019, Food and Agriculture Organization of the United Nations (FAO), Rome, Italy (García Préchac et al. 2019) and an invited conference in the 2019 Latin American Conference of Soil Science. The first two events were organized by the FAO Global Soil Partnership Organization and its Intergovernmental Technical Panel on Soils (ITPS), and the last by the Latin American Society of Soil Science.

2.4.1 THE URUGUAYAN OFFICIAL SOIL CONSERVATION POLICY

The Uruguayan law is based on the principle that soil conservation is of general interest, and according to the country's Constitution, it is dominant over any other particular interest. Therefore, regulation can limit what private landowners and land tenants do with soils over which they have control. The law determines that the Ministry of Agriculture is the authority that dictates the Technical Normative on soil conservation and prosecutes its fulfillment. The Technical Normative must be applied by the land tenants, independent of the contract type or form of land tenancy giving the tenant's right of land use. This Technical Normative is established by executive resolutions of the authority. The last law modification in 2009 added that when the land tenant is not the landowner, the latter is jointly liable with the tenant in case of regulatory violation. This provision incentivized the landowners to be the first stewards of soil conservation and was triggered by the majority of the country cropping being conducted by land renters and not landowners. The violation fines can range from US$300 to US$300,000, depending on their gravity.

Several general technical norms define "bad" soil use and management practices, like the generation of oriented roughness parallel to the main slopes, performing tillage operations or herbicides applications in areas of surface runoff concentration leading to gully erosion, etc. However, the main technical norm from 2009 is that each unit of soil management (lot or parcel) should file a Soil Use and Management Plan (SUMP) for the immediate future and duration of the planned rotation. The SUMP information contains (1) precise geographic location, (2) description of the rotation to be used, including all soil management details, (3) projected yields of the different crops, and eventually, pastures in the rotation, (4) dominant soil in the polygon, identifying to which Official Soil Map (1:1 M) unit the parcel belongs and (5) topographic characteristics defining L and S in the USLE/RUSLE. This model was validated in the country in long-term experiments (Pérez-Bidegain et al. 2018). Using the identified information, an annual average soil erosion rate is estimated with USLE/RUSLE, using EROSION 6.0.20 (García Préchac et al. 2016), and must be presented to the authority demonstrating that it is below the tolerance officially established for the particular soil.

A certified agronomist, contracted by the land tenant, prepares the technical report. To be certified, the agronomist must pass a specific exam given by the Faculty of Agronomy of the public University of Uruguay, working in agreement with the Ministry of Agriculture. The professional work of the certified agronomists is reviewed by an ethical and technical board.

Once the SUMPs are presented, the General Direction of Natural Resources of the Ministry of Agriculture studies and verifies their implementation, using satellite images and other forms of remote sensing, visiting the field when irregularities are suspected and contacting the responsible agronomists and farmers. If technical irregularities are found in a SUMP, the authority suspends the responsible agronomist certification. If irregularities are detected in the implementation, the land tenants (and eventually the landowners) can be fined, as previously mentioned.

Some key points to be highlighted are the following. A general law identifies the executive authority and the ones subject to prosecution. The authority can change the Technical Normative of soil conservation when needed via executive action. The procedure is based on punishment for the violations (regulations) and does not use monetary incentives. This system adapts to a variety of changes the country might experience, including a climate more degrading to soils. A regional survey comparing trends and means of the period 1931–1960 vs. 1971–2000, found a 15–20% increase in precipitation, mainly during the summer. The countries of the La Plata River Basin, through its coordinator committee, developed a program for the strategic planning of the water resources management considering climate change; one of the studies focused on estimating future precipitation impact on the rainfall erosivity R factor of USLE/RUSLE (Soares 2015). Even though this modeling exercise was based on one regional-scale model and only one RCP scenario (4.5), a 10% increase in rainfall erosivity is to be expected over the next 20 years. This change can be incorporated in the modeling systems used to evaluate farm management conservation performance.

Before requiring SUMPs in 2013, there were three years of extension work and training for farmers and agronomists. It consisted of short courses, workshops, and field work on selected cases, including the study of antecedent information prior to visiting the field and evaluating the SUMPs. This was a critical stage in the process, contributing to the results that are presented below.

2.4.1.1   Policy Results: Erosion Reduction

By February 2019, around 96% (14,000 km^2) of the Uruguayan cropland for which a SUMP presentation and implementation was required had been filed with the official authority. In February 2019, 16,840 SUMPs were filed; 1,625 of them had field visits. In 941 cases (5.6%) normative violations were found. This is considered very successful relative to mitigating soil erosion and collateral environmental impacts caused by runoff.

Finally, the most important issue is that the required SUMPs are evaluated before use in the field, meaning the policy has capacity to prevent use of soil-degrading practices. As discussed, this approach ensures acceptable soil use and management practices based on available science and is adaptable to a changing climate. This science-based proactive approach ensures Uruguay's soil resource will be preserved. This approach was implemented two years before publication of the UN Sustainable Development Goals and the Voluntary Guidelines for Sustainable Soil Management (FAO 2017); the Uruguayan soil conservation approach resulted in the application of "best management practices" (see Pérez-Bidegain et al. 2018), which are the foundation of achieving neutral land degradation and sustainable soil management.

2.4.1.2   Policy Results: Soils Organic Carbon (SOC) Content

Three long-term experiments in Uruguayan Argiudols, comparing soil use and management alternatives, have been used to verify modeling predictions of different management systems impact on SOC. Their results indicate the practices that reduce soil erosion also favorably affect SOC. The oldest experiment started in 1962 in the Experimental Station INIA-La Estanzuela ($34°20'12.31''$ S and $57°41'08.14''$ W). Soil is a Typic Argiudol, silty clay loam, with an original SOC of 2.2% in the 0–20 cm depth; site slope is 3.5%. Results during the period of conventional tillage soil management were reviewed by García Préchac et al. (2004, 2017). The second experiment started in 1994 in the Experimental Station EEMAC, Faculty of Agronomy-University of Uruguay ($32°2340.13$ S and $58°03'26.48$ W). The soil is a Typic Argiudol, clay loam, with an original 3% gravimetric SOC. The site differs from La Estanzuela in that its slope is less than 1%. Treatments are CC or CPR (3 years crops–3 years pasture), combined with conventional tillage or NT. SOC content was determined at the 0–15 cm depth after one rotation cycle of the CPR and discussed in two reviews (García Préchac et al. 2004, 2017). The third experiment is in the Experimental Unit INIA-Palo a Pique ($33°20'12.31''$ S and $54°29'34.43''$ W), in an Abruptic Argiudol, silty loam, with an original SOC of 1.7% in 1995. The experimental area is 72 ha, with 6 ha experimental units (Terra and García Préchac 2001). All soil is managed with NT. Soil uses contrasted are (1) CC: annual winter oats and ryegrass directly grazed, and sorghum or moha in summer for silage or hay; (2) SR (short rotation): two years idem CC and two years pasture; (3) LR (long rotation): two years idem CC and four years pasture; and (4) PP (permanent pasture): regenerated natural pasture over seeded with perennial legumes. After eight years, SOC and particulate organic carbon (C-POM, 53–2,000 μm) were determined at the 0–15 cm depth (Terra et al. 2006). Results from these long-term experimental results confirmed modeled predictions for most management trials across locations, thus providing a tool for determining cropping management system impacts on the soil carbon component of soil sustainability.

2.4.2   Crop–Pasture Rotations – Resilient Agro-ecological Production Systems for Uruguay

There are several definitions of *agroecology*, but all of them share the following elements: diversity, synergy, efficiency, recycling, resilience, co-innovation, human and social values, food culture and traditions, responsible governance, and solidarity and circular economy. Rotations of crops and pastures comply with most elements of agroecology, in addition to minimizing soil erosion and maintaining or increasing SOC. Crop–pasture rotations include plant and animal production, with variants for each, making them *diverse*. Symbiotic nitrogen fixation by pasture legumes increases nitrogen availability for the following crops, which is a *synergistic* element, as well as breaking weed, pest and disease cycles during the pasture period. These factors coupled with recycling management of nutrients such as phosphorous provide *efficiency*. Grazed pastures *recycle* potassium; this nutrient is absorbed by roots of perennial pastures, generally deeper rooted than those of row crops, and during grazing, it is returned to the surface in animal droppings, and therefore not exported in the animal product. The greater SOC content in rotations, apart from improving soil physical properties and nitrogen availability, also favorably affects sulfur availability, a major nutrient provided by soil organic matter and a nutrient that has begun to show a response in continuous cropping systems. Because of their diversity, rotation systems are more *resilient* to changes due to climate or economic conditions associated with input prices or product prices. Crop–pasture rotations have been the product of research, transfer, and adoption by producers since mid-1960s. They had been the predominant production systems before the extraordinary soybean production increase at the beginning of the present century; that is, they are the product of *co-innovation*.

Commercial inputs in CPRs compared to CC are much lower (Ernst and Siri-Prieto 2009). No-till rotations, where pastures occupied 40% of the time and area, reduced use of nitrogen fertilizer, phosphate fertilizers, glyphosate, other herbicides, fuel and machinery field time by 44%, 46%, 42%, 50%, 48% and 48%, respectively. Essentially no inputs are needed during the time pastures are present in the rotations. The combination of resiliency and reduced inputs further clarify the importance of this system in the face of a changing climate.

2.5 LOESS PLATEAU REGION OF CHINA: CLIMATE CHALLENGE(S) AND ADOPTION STRATEGIES

The Loess Plateau region is located at the midpoint of the Yellow River in China (latitude 34°–40° N and longitude 110°–115°), covers an area of 640,000 km^2 and is globally unique because of its large area and depth of loess. It is a semiarid area with annual precipitation ranging from about 150–750 mm, with an average of approximately 400 mm. The distinct seasonality of precipitation results in most of the rainfall occurring from June to September. Erosion in this region is one of the most significant environmental problems in China; it contributes 90% of the Yellow River sediment, making the Yellow River one of the world's most sediment-laden river. In the last century, the Loess Plateau has been losing an average of 1 cm of soil each year. The environment here is fragile.

Agriculture in the Loess Plateau region developed early in its history, beginning about 6,000 years ago. This area is one of the world's birthplaces of agricultural civilization. Soil erosion severity has increased in this region in the last half of the

20th century because of rapid population growth and the need for agricultural land; unfortunately, this increase has occurred with lack of proper management. Climate change, heat and drought covering this region, and increasing intensive precipitation events in some locations are the two most important future challenges (Sun et al. 2019; Zhao et al. 2018). These conditions threaten agriculture development and elevate the risk for intensive soil erosion events.

The Chinese government has prioritized soil and water conservation and agriculture in this region. A series of practices have been implemented and resulted in significant improvements in both environment and agriculture since the late last century. The sediment load of the Yellow River decreased by about 90% in the past 60 years (Wang et al. 2016) and grain self-sufficiency increased to 105.25% (Shi et al. 2020). In the following paragraphs, we describe three of the most useful practices in this area, including the "Grain for Green" program, terrace and check dam construction, and the integrated management of small watersheds.

2.5.1 "GRAIN FOR GREEN" PROGRAM

Aimed at both preventing land degradation and controlling Yellow River sediment, a series of afforestation programs in the Loess Plateau region and neighboring areas of China were implemented, beginning in the mid-20th century, e.g., the "Three North" Shelterbelt Development Program (TNSDP), the Natural Forest Conservation Program (NFCP) and the "Grain for Green" program (GGP) (Zhang et al. 2016). Among these programs, GGP is recognized as the largest ecological restoration and rural development program in the world (Delang and Yuan 2016), which involves 16 million farmers in 20 provinces (Wang et al. 2014). Farming steep slopes and overgrazing were two of the most important causes of land degradation in the Loess Plateau region. By GGP, steep sloping croplands (slopes of 25° or greater) were converted to native vegetation (tress or grassland, based on the water balance), thus minimizing or eliminating soil erosion and protecting the land and environment. The government compensated farmers for their loss of agricultural land.

The GGP has been very successful since its implementation in 1999 by the Chinese government. From 1999 to 2008, 33.3 million ha of land in the region was returned to forest and grasslands (including hills closed for afforestation) and the average vegetation coverage increased from 31.6% to 63.6% (Shan and Xu 2019). The greatest increase in vegetation coverage was found in the northern, central and western Loess Plateau (Li et al. 2019), where soil erosion is most severe. Vegetation restoration driven by this land-use change significantly improved the ecological environment, which helped reduce floods and sediment transport. Soil carbon loss on the Loess Plateau was also reduced since it is mainly attributed to sediment transfer (99.6%) (Deng et al. 2019).

At the beginning of the GGP, people worried whether the large areas of cropland converted to native vegetation would threaten the food supply. Indeed, grain production fell in the first three years of the GGP from 1999 to 2001. In the subsequent years, GGP was combined with improving comprehensive agricultural production capacity and developing high-quality farmland. Taking Wuqi county in the Loess

Plateau as an example, 15,200 ha of new basic farmland was developed since 2003 resulting in crop yield increases from 870 to 2,820 kg·ha⁻¹ from 1998 to 2014 (Wang et al. 2014). This agriculture success was gained with increased agricultural material input and construction of terraces and check dams to ensure food supply in this area is sustained (Shi et al. 2020), which is discussed in detail in the following section.

2.5.2 TERRACE AND CHECK DAM CONSTRUCTION

The construction of terraces and check dams is primarily responsible for the sediment load reduction of the Yellow River (Wang et al. 2016). This practice also improves food security through creating fertile flat agricultural land (Shi et al. 2020) in the Loess Plateau region. By 2018, 3.36 million ha of terraces had been built in the Loess Plateau region (Gao et al. 2020). Terraces increase rainfall infiltration and reduce surface run-off and soil loss, thus increasing crop yield and ultimately productivity. Check dam use in the Loess Plateau region began at least 400 years ago. In the past century, the two most important periods for check dam construction were 1968–1976 and 2004–2008. Prior to 2015, 56,422 check dams above the Tongguan station were constructed (Liu, X., Gao, Y., Ma, S., & Dong, G. 2018). Check dams, temporary structures designed across drainage channels, are effective measures for stopping coarse sediment transport to the Yellow River and was reported to be the dominant approach for sediment control in sub-catchments of this region (Da-Chuan et al. 2008). The agricultural landscape has been changed from sloping cropland to terraced and check dam–dominated farmland in many areas of the Loess Plateau region.

With climate change and more intense precipitation, management of terrace and check dams is becoming a daunting challenge in some areas of the Loess Plateau region. Terraces and check dams can be seriously damaged by extreme rainstorms if not well managed, causing huge sediment loads from the damaged terrace and check dam farmlands. The efficiency of terraces and check dams in controlling soil loss was greatly reduced under extreme rainstorms. For example, in an extreme rainstorm of July 26, 2017, the sediment interception efficiency of the check dams was only 26.36% in one of the watersheds in the Loess Plateau, and most of the sediment was transported to downstream channels (Bai et al. 2020). The Chinese government is now paying more attention to improving the management and increasing the construction technology standards of terraces and check dams for both sediment control and flood safety. Experience indicates that it is important to combine these practices with other kinds of conservation measures to obtain satisfactory results, an approach called integrated management of small watersheds.

2.5.3 INTEGRATED MANAGEMENT OF SMALL WATERSHEDS

Comprehensive prevention and control systems that combine engineering, vegetation and agriculture technology measures in a watershed are proving to be more effective in soil loss control than single measures in many areas of the world. Management models for watersheds vary with different geographical factors and human activities. In the Loess Plateau region, this integrated management in small watersheds began in the 1950s with the establishment of five management modes based on experiments in 11 typical small watersheds. The five models include the eco-agriculture model, wood and grass model, traditional agriculture model, economic forest model and agroforestry model. These models conserve soil; soil loss reductions of 50–90% were observed after ten years of the practice (Liu, G., Yang, Q., & Zheng, F. 2004). After more than 20 years, this approach developed into a more comprehensive and systematic model and began to be applied in regional-scale management. In the 1980s, the concept of "small watershed integrated management" was introduced by Chinese scientists, taking a small watershed with a drainage area of 4–50 km^2 as a unit, and combining soil conservation measures inside the watershed. By 2012, more than 50,000 watersheds were enrolled in this integrated management practice in China and this became the primary approach for both environment and farmer income improvement in the Loess Plateau region and across China.

A scientist named Zhu Xianmo, who is one of the most famous and important soil conservation scientists in China, summarized the integrated management of small watersheds in the Loess Plateau region in 28 Chinese characters: "全部降水就地入渗拦蓄，米粮下川上塬、林果下沟上岔、草罐上坡下坬" (Zhu 1991). This means:

> in the Loess Plateau we should try our best to have the precipitation infiltrate and stored, crops could be down at the bottom of the river valleys or go up to the plain plateau area, orchard should be planted in gully areas where water is not so limited, while on steep slopes especially without much water storage, grass and shrub should be the right choice.

After nearly 70 years, integrated management is playing a more and complete role in soil loss prevention. Even in extreme rainstorms, watersheds with well-integrated management schemes have avoided serious soil loss–related disasters. For example, for the 100-year return rainstorm of July 26, 2017, most of the check dams were undamaged and continue playing important roles in stopping sediment transport into the Yellow River from the Jiuyuan watershed, which is located in the most serious erosion area of the Loess Plateau. This success is attributed to grassland conversion across most of the steeply sloping croplands, terrace construction on many gentle slopes and check dam construction along the entire length of streams (Liu, B., Liu, X., Yang, Q., Zhang, X., Cao W., & Dang, W. 2017).

2.6   NE CHINA MOLLISOL (BLACK SOIL REGION): CLIMATE CHALLENGE(S) AND ADOPTION STRATEGIES

2.6.1   Brief Introduction

Black Soils of Northeast China, located between 38°43′–53°33′ N and 115°31′–135°05′ E, are considered the most fertile soil resource of China – perhaps the world (Lal, R. 2012). Black soils are of global importance because of their role in food security. Nearly one-third of China's commercial grains are provided by the Black Soil region. However, interactions of geography, climate and intensive farming have made the Black Soil region one of the most seriously degraded areas in China, threatening national food security and regional development. Black soil is described as China's giant panda of cultivated land; on the one hand, it is scarce and valuable and, on the other hand, it is vulnerable to human activity and climate change. For example, the geographic centroid of rice production shifted over 320 km northeastward during the past 30 years due to climate change (Hu et al. 2019). Crop yields may benefit from global warming due to effective accumulated temperature increases (Zhou et al. 2013); however, soil erosion rate has also accelerated due to more intense precipitation (Han et al. 2019; Liu, B., Xie, Y., Liang, Y., et al. 2020a). Additionally, the high soil organic carbon (SOC) content of black soils makes them a potentially large sources of greenhouse gases.

With production pressure associated with increasing food demand and concurrent global warming stresses, black soils must be managed appropriately to maintain their productivity and mitigate climate change. Undoubtedly, soil and water conservation measures are the top priority for adapting to a climate with more erosive rainfall events. Conservation tillage approaches, vegetation (e.g., shelterbelt, agroforestry) and engineering practices have been separately or jointly adopted to prevent soil degradation at different landscape scales in this area (Zhang and Li 2014; Deng et al. 2015). The combinations of agricultural and forestry practices (e.g., Mount ain-River-Forest-Farmland-Lake-Grass Protection and Restoration) are the leading approaches for effectively retaining soil/soil-associated carbon and concurrently capturing carbon in this landscape. As a consequence, climate-friendly agricultural practices in Northeast China will likely transition farmland from a greenhouse gas source to a carbon sink, which is an important component for China to achieve carbon neutrality by 2060.

2.6.2 CONSERVATION TILLAGE

In China, the adoption of conservation tillage (CT) practices has been actively encouraged since 2002 following the recognition of increased soil degradation rates due to water erosion in Northeast China (Wang et al. 2007). After years of systematic research and demonstration, the CT farming system has been trusted as a sustainable approach to reduce soil erosion, increase water infiltration, increase resilience to drought, reduce soil carbon loss and mitigate climate change.

CT, which varies in its practice among regions in China, has been defined by the Conservation Technology Information Center (CTIC) in China as a tillage system that leaves a crop residue cover of at least 30% after crop planting (Kuhwald et al. 2018). According to the CTIC, in China there are four main categories or types of CT practice (no-tillage, mulch tillage, reduced or minimum tillage and ridge tillage). Ridge tillage and no-tillage are commonly adopted in Northeast China, sometimes combined with changing ridge orientation and/or size. Compared with the traditional narrow longitudinal ridge tillage, the relative soil conservation associated with wide longitudinal ridge tillage decreases with increased precipitation. For example, when *PI30* (product of event precipitation *P* and maximum 30 min intensity *I30*) was around 500, the reduction of soil loss associated with wide longitudinal ridge tillage, compared to the narrow longitudinal ridge tillage, was as much as 90% and when PI30 was around 1,600, the reduction of soil loss by the wide longitudinal ridge tillage was as high as about 65%. While the relative effect varied with precipitation erosivity, wide longitudinal ridge tillage maintains better soil conservation effects than the narrow longitudinal ridge tillage system in the Black Soil region of China (Wang et al. 2019).

Compared with conventional tillage that leaves the field surface bare after harvest, chopping and returning 100% of the crop residue to the field surface can reduce soil loss more than 90% (Liu, H., Shan, Z., Qin, W., & Yin, Z. 2020b). Their field experiment found that subsoiling and returning crop residue and furrow damming worked more effectively than conventional tillage at reducing runoff and sediment losses – about 95% and 90%, respectively; meanwhile, the yield and the water use efficiency can be increased by about 25% and 30%, respectively, compared to conventional tillage (Wei et al. 2013). CT coupled with longer crop rotations could also positively affect crop yield and profitability. For instance, corn–soybean under NT produced better yield and profitability, particularly in dry years, than the three continuous years of corn or corn–corn–soybean rotations with conventional tillage (Wang and Zhang 2008).

Lishu County in the Black Soil region of Northeast China, one of the top five national grain production counties, is an eye-catching example of recent CT implementation and its soil conservation effects. During August–September 2020, a series of three typhoons – Bavi, Maysak and Haishen – brought heavy rain and strong wind to the Black Soil region within half a month, lodging crops and flooding fields. The average rainfall was as high as 170 mm, three times higher than the normal for the same period based on data since 1961. However, a 130,000 ha CT demonstration field of corn in this area still produced 13,500 kg·ha^{-1}, with a 10% yield increase compared to conventional tillage. Over 14 years of CT, the soil organic matter content increased 1.68%, the corn rooting depth increased to 1.2 m and more than 100 earthworms per m^2 were found in the plow layer, with earthworm tunnels extending 0.5 m deeper than observed for conventional tillage. Additionally, no-tillage promoted C accumulation within microaggregates which then form macroaggregates. The shift of SOC within microaggregates is beneficial for long-term C sequestration in the soil managed with CT (Zheng et al. 2018).

Given CT's noticeable ecological and economic contributions, many policies for CT have been launched by the Chinese government for the rapid expansion of CT, such as the latest Central Document issued in March 2020 by the Ministry of Finance and Ministry of Agriculture, "Conservation Tillage Action Plan for Black Soils of Northeast China (2020–2025)" (Ministry of Agriculture and Rural Affairs of the People's Republic of China (2020-2025), Publication date: 2020-03-19). A clear goal was put forward that CT systems will be extensively implemented in the Black Soil region with an area reaching as much as 933,300 km^2 by the end of 2025, accounting for 70% of the arable farmland of Northeast China.

2.6.3 Soil and Water Conservation Forests

Limited forest resources and a fragile environment make China's natural resources susceptible to soil erosion, water runoff, wind- and sandstorms, floods and droughts. To address these problems, China has implemented an afforestation program as a national policy and has made it one of the major components of ecological conservation and environmental restoration along with economic development (Zhang 2002). Several national forestry eco-engineering projects oriented toward soil and water conservation, environmental protection and forest resource expansion have been conducted since 1978. Of these, the Three-North Afforestation Program (TNAP), scheduled to be completed in 2050, is the world's largest afforestation project and covers 4.07×10^6 km^2 (42.4%), including Northeast, Northwest and North Central China (Zhu et al. 2017). As one part of the TNAP, dense farmland shelterbelt networks in northeastern China have reduced water and sediment flows. For instance, at the Keshan Farm, the mean soil erosion rate and specific sediment yield (defined as the ratio of soil sediment to catchment area [t·km^{-2}·yr^{-1}]) of the 25 reservoir catchments decreased from 351.6 and 93.9 t·km^{-2}·yr^{-1} under the modeled scenario without shelterbelts to 331.1 and 86.3 t·km^{-2}·yr^{-1}, respectively, with the current shelterbelts. The sediment-trapping efficiencies varied from 0.01% to 23.6% with an average value of 7.6% (Fang 2021).

The two shelterbelts (upper and lower) in the catchment reduced soil loss to a certain degree; sediment deposition usually occurred both above and below the shelterbelts, leading to less sediment delivery compared to that which existed before the shelterbelts were installed. Conversely, erosion became severe below the shelterbelts as filtered water flowing through the shelterbelt exited on the lower side. In comparison to the lower shelterbelt, the upper shelterbelt with more gentle topography and lower slope gradient between the shelterbelts had higher sediment-trapping efficiency. Impacted by the two lines of shelterbelts and the earth bank at the catchment outlet, the catchment resulted in an erosion–deposition–erosion–deposition pattern. In the recent 50 and 100 years based on model estimates, the sediment amounts annually deposited along the shelterbelts at distances of 30 m and 60 m contained 18.8% and 7% of the total deposited sediment in the catchment, respectively. The sediment delivery ratios derived by 137 Cs and 210 Pb methods were 53% and 78%, respectively (Fang and Wu 2018).

The shelterbelt system also plays an important role in the control of gully erosion directly or indirectly in rolling hills of the Black Soil region (Heshan Farm). With increasing slope gradient, there was an inverse trend between gully density and shelterbelt density, indicating that farmland shelterbelts can prevent gully erosion. The positive effect of farmland shelterbelts against gully erosion varied with distance: for distances <120 m from the shelterbelt, the effect was consistent and very strong; for distances of 120–240 m, a weak linear decrease was found in the gully-avoidance effect; and for distances >240 m, the effect of the shelterbelts was significantly weaker. A recommended optimal planting density of farmland shelterbelts for the prevention of gully erosion is 1,100–1,300 m·km^{-2} (Deng et al. 2015).

Shelterbelts increased maize yields 4.7%, 4.3% and 9.5% in the high, middle and low climatic potential productivity zones, respectively (Zhu et al. 2017). In Northeast China, on average 18.3% of the farmland is protected by shelterbelts, which is obviously lower than the optimal level of protection (i.e., approximately 80%), thus, many shelterbelts remain to be planted in the future (Zheng et al. 2016). It is important to note that some poorly designed windbreaks have accelerated soil loss.

Effective shelterbelts are mainly located in two geomorphology units: the convex position of the hill and the conjunction of two hill slopes. In the typical convex area, the new plan estimates that soil loss can be reduced 32.3%, while in the typical conjunction area of two hill slopes, an estimated 48.4% reduction in soil loss is expected (Su et al. 2013). The proposed scheme for optimizing the distribution of existing shelterbelts includes the adjustment of shelterbelt orientation such that they are perpendicular to the slope gradient, improved maintenance, planting of trees in shelterbelts to reduce open gaps, increase in shelterbelt number and decrease distances between shelterbelts (Zhu et al. 2017).

2.6.4 INTEGRATED SMALL WATERSHED MANAGEMENT

Soil and water conservation practices are one of the most important components of an integrated watershed program, particularly for the rolling hill area in the Black Soil region. A Three Defense Lines control pattern is most commonly used in this area. The first defense line is the protection system established on the upper portion of the slope. An interception ditch and farmland shelter forest are arranged along the hilltop and road, which retain water from the slope, thus avoiding inundation of farmland below. The second defense line is the protection system setup on the slope. Based on farmland slope gradient and length, appropriate methods are applied to conserve soil and water and to increase land productivity. If the slope is <3°, level ridge tillage is used on the farmland; if the slope is 3–5°, a vegetation belt is planted on the slope; if the slope is 5–7°, a level terrace is constructed on the slope; and if the slope >7°, returning farmland to forest and grassland is preferred. The third defense line involves a protection system in the gully position. The recommended measures for preventing gully expansion from devouring farmland include constructing flow cascades on the gully head, building check dams on the gully bed, reducing the slope of the gully wall and increasing gully vegetation, as well as terminating crop production on hillsides to facilitate afforestation (Sun et al. 2012).

Integrated small watershed management adheres to the principles of the Mountain-River-Forest-Farmland-Lake-Grass System of Ecological Protection and Rehabilitation Program, issued by the Ministry of Natural Resources, Ministry of Finance and Ministry of Ecological Environment in 2020. This is the first program to emphasize the Community of Shared Life in ecological protection. It is a guiding principle for soil conservation, ecological restoration and regional economic development. The concept of meta-ecosystem has been widely acknowledged by the farmers from Baiquan County of the Black Soil region, which was removed from the national-level poverty counties list in 2020. A household doggerel "Shi Zi Deng Ke" is used to vividly summarize the measures and achievements of integrated small watershed management:

> Pinus Sylvestris is planted on the slope top like a hat, Lespedeza bicolor is planted on the terraced ridge like a scarf, farmland is converted to grassland like a blanket, fish are raised in the ditch, ducks are raised within the dam, rice is cultivated outside the dam, fruit trees are planted in the swales, shelterbelts are planted on the flat, making

a united effort to open factories, and earn more money by comprehensive ecological management.

(Wang and Su 1995)

2.7   EUROPE: CLIMATE CHANGE AND SOIL EROSION

Globally, climate change is projected to increase severe storm intensity (Brooks 2013), increasing runoff and decreasing infiltration in arable crops (Basche and DeLonge 2017) which may cause even greater soil losses than that which occurred in the beginning of the 21st century (Borrelli et al. 2017). Therefore, climate change has been addressed in the last decade in relation to increased soil erosion losses. The recent devastating catastrophic floods in Germany, Belgium and Luxembourg (July 8, 2021) show that climate is changing faster than expected and is affecting mostly the northern part of Europe with higher intensity rainfalls. However, the excessive heat-wave over Greece and South Italy with temperatures exceeding 45°C for more than ten days (July 27–August 6, 2021) is another example of accelerated climate change on the European continent. In Europe, selected studies have addressed, and others are addressing, the impact of climate change–driven intense rainfall on increased erosion. Among others, we reference certain studies in Belgium (Mullan et al. 2019), Greece (Grillakis et al. 2020) and Austria (Luetzenburg et al. 2020).

Recently, important developments in climate change data addressing rainfall projections for the 2041–2060 period at highest requested spatial resolution (30 second) became available in the WorldClim database (Fick and Hijmans 2017). These data were developed based on a large variety of 19 Global Climate Models (GCMs) across three Representative Concentration Pathways (RCPs). In addition, the Rainfall Erosivity Database at the European Scale (REDES) includes high temporal (hourly, sub-hourly) precipitation data and long-term erosivity values (Panagos et al. 2015) and it allows for the development of Gaussian Process Regression (GPR) models (Williams and Rasmussen 2006). The GPR model establishes a statistical relationship between the actual erosivity values of REDES and the rainfall records in WorldClim.

Compared to the first decade of the 21st century, the rainfall erosivity is expected to increase in the range of 22–37%, depending on the RCP scenario. Those are average values, of the 19 climate model applications, for the whole European Union and the United Kingdom. In the most aggressive mitigation pathway scenario RCP2.6, the 22% increase of erosivity will result in a mean soil loss increase of 13% in the EU and UK with three-fourth of the continent showing increasing trends while the rest a slight decrease (Panagos et al. 2021). Similar to the RCP2.6, the RCP4.5 will have the same trends and mean increases but somehow some different spatial patterns. Finally, the RCP8.5 projects a mean increase of rainfall erosivity of about 37%, resulting in increase of soil losses close to 26% (Figure 2.4). Therefore, the mean soil loss in the agricultural lands of the baseline (2016) is about 3.07 t·ha^l·yr^l (Panagos et al. 2020) and it is expected to increase to 3.76 t·ha^l·yr^l in the business as usual or least mitigation pathway scenario RCP8.5 in 2050. The results are in line with global projections of soil erosion developed by Borrelli et al. (2020).

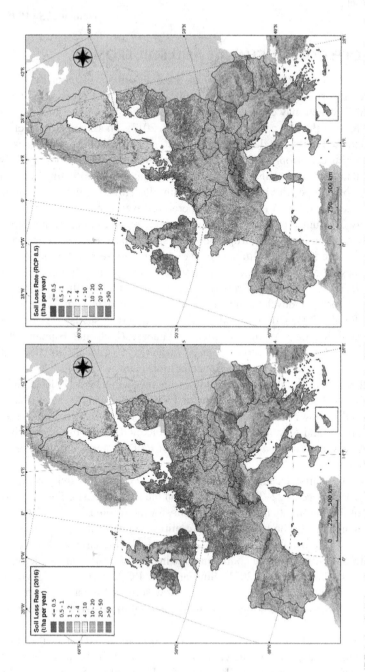

FIGURE 2.4 Soil loss by water erosion in agricultural lands of the EU and UK for the baseline period (2016) and the future projections for 2050 (RCP8.5 scenario). Source: Created by Panos Panagos.

2.7.1 Mitigation Measures

Projected land use changes (conversion of arable lands to pastures) mitigate a small part of the climate change impacts on soil erosion. As the share of pastures will increase by 2% by 2050 and the share of arable lands will decrease by 2.1%, this may have a mitigation effect of decreasing soil losses up to 3% in the EU. This small portion needs to be reinforced with mitigation measures applied through agro-environmental policies in the EU. The most effective policy instrument is to link Common Agricultural Policy incentives to famers with their environmental performance in a targeted way (Panagos and Katsoyiannis 2019). However, the application of soil conservation measures such as cover crops and reduced tillage should focus in hotspots and should include at least 50% of the area with soil loss rates higher than 5 t·ha^1·yr^1. Such an aggressive soil conservation policy may neutralize the future impact of climate change on water erosion.

LITERATURE CITED

Amundson, R., Asefaw Berhe, A., Hopmans, J. W., Olson, C., Ester Sztein, A., & Sparks, D. L. (2015). Soil and human security in the 21st century. *Science*, *348*(6235), 1261071. doi: 10.1126/science.1261071.

Arbeletche, P., Ernst, O., & Hoffman, E. (2010). La Agricultura en el Uruguay y su evolución. In G. Préchac et al. (Eds.), *Intensificación Agrícola: Oportunidades y amenazas para un país productivo y natural* (pp. 13–27). Uruguay: CSIC-Universidad de la República.

Bai, L., Wang, N., Jiao, J., Chen, Y., Tang, B., Wang, H., ... Wang, Z. (2020). Soil erosion and sediment interception by check dams in a watershed for an extreme rainstorm on the Loess Plateau, China. *International Journal of Sediment Research*, *35*(4), 408–416.

Basche, A. & DeLonge, M. (2017). The impact of continuous living cover on soil hydrologic properties: A meta-analysis. *Soil Science Society of America Journal*, *81*(5), 1179–1190.

Ben-Gal, A., Karlberg, L., Jansson, P. E., & Shani, U. (2003). Temporal robustness of linear relationships between production and transpiration. *Plant and Soil*, *251*(2), 211–218.

Beretta-Blanco, A., Pérez, O., & Carrasco-Letelier, L. (2019). Soil quality decrease over 13 years of agricultural production. *Nutrient Cycling in Agroecosystems*, *114*(1), 45–55.

Borrelli, P., Robinson, D. A., Fleischer, L. R., Lugato, E., Ballabio, C., Alewell, C., Meusburger, K., Modugno, S., Schütt, B., Ferro, V., Bagarello, V., Van Oost, K., Montanarella L., & Panagos, P. (2017). An assessment of the global impact of 21st century land use change on soil erosion. *Nature Communications*, *8*(1), 1–13. doi: 10.1038/s41467-017-02142-7.

Borrelli, P., Robinson, D. A., Panagos, P., Lugato, E., Yang, J. E., Alewell, C., Borrelli, P., Robinson, D., Panagos, P., Lugato, E., Yang, J., Alewell, C., Wuepper, D., Montanarella, L. & Ballabio, C. (2020). Land use and climate change impacts on global soil erosion by water (2015–2070). *Proceedings of the National Academy of Sciences*, *117*(36), 21994–22001.

Brooks, H. E. (2013). Severe thunderstorms and climate change. *Atmospheric Research*, *123*, 129–138.

Buda, A. (2021, August 2). The role of climate-smart agriculture in climate adaptation and mitigation in the Northeast. Retrieved from https://www.climatehubs.usda.gov/hubs/northeast/topic/role-climate-smart-agriculture-climate-adaptation-and-mitigation-northeast.

Cayssials, R., Liesegang, J., & Piñeyrúa, J. (1978). Panorama de la erosión y conservación de suelos en Uruguay. In Dir. de Suelos y Fertilizantes-MAP, *Boletín Técnico No. 4.*

Da-Chuan, R., Quan-Hua, L., Zu-Hao, Z., Guo-Qing, W., & Zhang, X. H. (2008). Sediment retention by check dams in the Hekouzhen-Longmen Section of the Yellow River. *International Journal of Sediment Research, 23*, 59–166.

Delang, C. O., & Yuan, Z. (2016). *China's grain for green program*. Berlin: Springer, 230 p.

den Biggelaar, C., Lal, R., Wiebe, K., & Breneman, V. (2003). The global impact of soil erosion on productivity: I: Absolute and relative erosion-induced yield losses. *Advances in Agronomy, 81*, 1–48.

Deng, L., Kim, D. G., Li, M., Huang, C., Liu, Q., Cheng, M., Peng, C. (2019). Land-use changes driven by 'Grain for Green'program reduced carbon loss induced by soil erosion on the Loess Plateau of China. *Global and Planetary Change, 177*, 101–115.

Deng, R., Wang, W., Fang, H., & Yao, Z. (2015). Effect of farmland shelterbelts on gully erosion in the black soil region of Northeast China. *Journal of Forestry Research, 26*(7), 941–948. doi: 10.1007/s11676-015-0110-4.

Durán, A. (1998). Contenido y distribución geográfica de carbono orgánico en suelos del Uruguay. *Agrociencia (Uruguay), 2*(1), 37–47.

Durán, A., & García Préchac, F. (2007). *Suelos del Uruguay, origen, clasificación, manejo y conservación*, Vol. 2, 357 p. Ed. Hemisferio Sur. Montevideo, Uruguay.

Ernst, O., & Siri-Prieto, G. (2009). Impact of perennial pasture and tillage systems on carbon input and soil quality indicators. *Soil and Tillage Research, 105*(2), 260–268.

Eswaran, H., Van Den Berg, E. & Reich, P. (1993). Organic carbon in soils of the world. *Soil Science Society of America Journal, 57*(1), 192–194.

Eswaran, H., Van Den Berg, E., Reich, P., & Kimble, J. (1995). Global soil carbon resources. In R. Lal, J. Kimble, E. Levine & B. A. Stewart (Eds.), *Soils and global change. Advances in soil science*. CRC Press Inc. Lewis Publishers - imprint of CRC Press, Boca Raton, FL 440 p.

Fang, H. (2021). Quantifying farmland shelterbelt impacts on catchment soil erosion and sediment yield for the black soil region, northeastern China. *Soil Use and Management, 37*(1), 181–195. doi: 10.1111/sum.12591.

Fang, H., & Wu, D. (2018). Impact of agricultural shelterbelt on soil erosion and sediment deposition at catchment scale in the black soil region northeastern China. *Journal of Shanxi Normal University (Natural Science Edition), 46*(1), 104110. (in Chinese with English abstract). doi: 10.15983/j.cnki.jsnu.2018.01.412.

Feng, Z., Ruby Leung, L., Hagos, S., Houze, R. A., Burleyson, C. D., & Balaguru, K. (2016). More frequent intense and long-lived storms dominate the springtime trend in central US rainfall. *Nature Communications, 7*, 13429. doi: 10.1038/ncomms13429.

Fick, S. E., & Hijmans, R. J. (2017). WorldClim 2: New 1-km spatial resolution climate surfaces for global land areas. *International Journal of Climatology, 37*(12), 4302–4315.

Fischer, G., van Velthuizen, H., Shah, M. and F. Nachtergalele. (2002). *Global agro-ecological assessment for agriculture in the 21st century: Methodology and results*. Research Report RR-02-02. Laxenburg: International Institute for Applied Systems Analysis. ISBN 3-7045-0141-7.

Food and Agriculture Organization of the United Nations. (2010). *"Climate-Smart" agriculture policies, practices and financing for food security, adaptation and mitigation*. Rome, Italy. Retrieved from http://www.fao.org/3/i1881e/i1881e00.pdf.

Food and Agriculture Organization of the United Nations. (2017). *Voluntary guidelines for sustainable soil management*. Rome: United Nations Food and Agriculture Organization.

Gao, Y., Wang, L., Wang, H., Li, P., & Wang, T. (2020). Thoughts on the construction of dry farming terraces in the Loess Plateau in the new era. *Soil and Water Conservation in China, 9*, 73–75.

García Préchac, F. (2004). Cultivo continuo en siembra directa o rotaciones de cultivos y pasturas en suelos pesados del Uruguay. *Revista Científica Agropecuaria, 8*(1), 23–29.

García Préchac, F., Clérici, C., & Hill, M. (2019). The Uruguayan official soil conservation policy and its results. In *Proceedings of the global symposium on soil erosion* (pp. 548–552). Rome: Food and Agriculture Organization.

García Préchac, F., Clérici, C., Hill, M., & Hill, E. (2016). EROSION v. 6.0.20 (computer program to use USLE/RUSLE in the southern La Plata river basin), Retrieved from www\fagro.Dpto.de Suelos y Aguas, Manejo y Conservación. Retrieved from edu.uy.

García Préchac, F., Ernst, O., Siri-Prieto, G., Salvo, L., Quincke, A., & Terra, J. A. (2017). Long-term effect of different agricultural soil use and management systems on the organic carbon content of Uruguay prairie soils. In *Proceedings of the global symposium on soil organic carbon* (pp. 449–452). Rome: Food and Agriculture Organization.

García Préchac, F., Ernst, O., Siri-Prieto, G., & Terra, J. A. (2004). Integrating no-till into crop pasture rotations in Uruguay. *Soil and Tillage Research, 77*(1), 1–13.

Gelder, B., Sklenar, T., James, D., Herzmann, D., Cruse, R. M., Gesch, K., & Laflen, J. (2018). The daily erosion project: Daily estimates of water runoff, soil detachment, and erosion. *Earth Surface Processes and Landforms, 43*(5), 1105–1117. doi: 10.1002/esp.4286.

Grillakis, M. G., Polykretis, C., & Alexakis, D. D. (2020). Past and projected climate change impacts on rainfall erosivity: Advancing our knowledge for the eastern Mediterranean island of Crete. *CATENA, 193*, 104625.

Han, T., Wang, H., Hao, X., & Li, S. (2019). Seasonal prediction of midsummer extreme precipitation days over Northeast China. *Journal of Applied Meteorology and Climatology, 58*(9), 2033–2048. doi: 10.1175/JAMC-D-18-0253.1.

Hu, Y., Fan, L., Liu, Z., Qiangyi, Y., Liang, S., Chen, S., You L., Wu, W., & Yang, P. (2019). Rice production and climate change in Northeast China: Evidence of adaptation through land use shifts. *Environmental Research Letters, 14*(2), 024014. doi: 10.1088/1748-9326/aafa55.

IPCC. (2019, revised in 2020). Climate change and land: An IPCC special report on climate change, desertification, land degradation, sustainable land management, food security, and greenhouse gas fluxes in terrestrial ecosystems. Contributions of working group to the assessment report of the intergovernmental panel on climate change. Cambridge: Cambridge University Press.

Kuhwald, M., Dörnhöfer, K., Oppelt, N., & Duttmann, R. (2018). Spatially explicit soil compaction risk assessment of Arable soils at regional scale: The SaSCiA-Model. *Sustainability, 10*(5), 1618. doi: 10.3390/su10051618.

Lal, R. (2012). Climate change and soil degradation mitigation by sustainable management of soils and other natural resources. *Agriculture Research, 1*(3), 199–212. doi: 10.1007/s40003-012-0031-9.

Li, G., Sun, J., Han, J. Y., Liu, W., Wei, Y., & Lu, N. (2019). Impacts of Chinese grain for green program and climate change on vegetation in the Loess Plateau during 1982–2015. *Science of the Total Environment, 660*, 177–187.

Liu, B., Liu, X., Yang, Q., Zhang, X., Cao, W., & Dang, W. (2017). Report on the efficiency of integrated soil conservation measures during extreme rainstorm in small watersheds in Loess Plateau. *Bulletin of Soil and Water Conservation, 37*(2), 349–350.

Liu, B., Xie, Y., Li, Z., Liang, Y., Zhang, W., Fu, S., … Guo, Q. (2020a). The assessment of soil loss by water erosion in China. *International Soil and Water Conservation Research, 8*(4), 430–439. doi: 10.1016/j.iswcr.2020.07.002.

Liu, G., Yang, Q., & Zheng, F. (2004). Small watershed management and eco-rehabilitation on the Loess Plateau of China. *Science of Soil and Water Conservation, 68,* 85–94.

Liu, H., Shan, Z., Qin, W., & Yin, Z. (2020b). Review on soil and water loss control techniques and models in the Black Soil regions in the Northeast. *Journal of Sediment Research, 45*(4), 74–80. (in Chinese with English abstract). doi: 10.16239/j.cnki.0468-155x.2020.04.012.

Liu, X., Gao, Y., Ma, S., & Dong, G. (2018). Sediment reduction of warping dams and its timeliness in the Loess Plateau. *Journal of Hydraulic Engineering, 49,* 145–155.

Luetzenburg, G., Bittner, M. J., Calsamiglia, A., Renschler, C. S., Estrany, J., & Poeppl, R. (2020). Climate and land use change effects on soil erosion in two small agricultural catchment systems Fugnitz–Austria, Can Revull–Spain. *Science of the Total Environment, 704,* 135389.

Ministry of Agriculture and Rural Affairs of the People's Republic of China. (2020–2025). Promoting the high-quality development of conservation tillage. Retrieved from http://www.moa.gov.cn/xw/bmdt/202003/t20200325_6339873.htm

Ministry of Agriculture and Rural Affairs of the People's Republic of China. *Conservation Tillage Action Plan for Black Soils of Northeast China,* (2020-2025). Publication date: 2020-03-19, Nanli, Chaoyang District, Beijing China (100125).

Morón, A. (2009). Estimaciones del Impacto de la Agricultura y la Ganadería en el Suelo en Uruguay. *El efecto de la agricultura en la calidad de los suelos y fertilización de cultivos, INIA, Serie Activ. de Dif, 605,* 15–18.

Mullan, D., Matthews, T., Vandaele, K., Barr, I. D., Swindles, G. T., Meneely, J., … Murphy, C. (2019). Climate impacts on soil erosion and muddy flooding at 1.5 versus 2° C warming. *Land Degradation and Development, 30*(1), 94–108.

Panagos, P., Ballabio, C., Borrelli, P., Meusburger, K., Klik, A., Rousseva, S. (2015). Rainfall erosivity in Europe. *Science of the Total Environment, 511,* 801–814.

Panagos, P., Ballabio, C., Himics, M., Scarpa, S., Matthews, F., Bogonos, M., Poesen, J., & Borrelli, P. (2021). Projections of soil loss by water erosion in Europe by 2050. *Environmental Science and Policy, 124,* 380–392.

Panagos, P., Ballabio, C., Poesen, J., Lugato, E., Scarpa, S., Montanarella, L., & Borrelli, P. (2020). A soil erosion indicator for supporting agricultural, environmental and climate policies in the European Union. *Remote Sensing, 12*(9), 1365.

Panagos, P., & Katsoyiannis, A. (2019). Soil erosion modelling: The new challenges as the result of policy developments in Europe. *Environmental Research, 172,* 470–474.

Paruelo, J. M., Jobbágy, E. G., & Sala, O. E. (2001). Current distribution of ecosystem functional types in temperate South America. *Ecosystems, 4*(7), 683–698.

Pérez-Bidegain, M., Hill, M., Clérici, C., Terra, J. A., Sawchik, J., & García Préchac, F. (2018). Regulatory utilization of USLE/RUSLE erosion rate estimates in Uruguay: A policy coincident with the UN sustainable development goals. In R. Lal et al. (Eds.), *Soil and sustainable development goals* (pp. 82–91). Stuttgart: Catena-Schweizerbart.

Sganga, J. C., Víctora, C. D., & Cayssials, R. (2005). *Plan de acción nacional de lucha contra la desertificación y la sequía.* ROU, DINARA-MOTVMA, RENARE-MGAP, Proy. GM2/020/CCD, 168 p.

Shan, L., & Xu, B. (2019). Discussion on some issues about returning farmland to forest or grassland on Loess Plateau in new era. *Bulletin of Soil and Water Conservation, 39,* 295–297.

Shi, P., Feng, Z., Gao, H., Li, P., Zhang, X., Zhu, T., Li, Z., Xu, G., Ren, Z., & Xiao, L. (2020). Has "Grain for Green" threaten food security on the Loess Plateau of China? *Ecosystem Health and Sustainability, 6*(1), 1709560.

Soares, W. R. (2015). Estimativas de erosividade a partir de simulações climáticas para a Bacia do Prata utilizando o modelo eta com resolução horizontal de 20 km: Projeções Futuras, anomalias e médias móveis. Componente III: Modelos Hidro climáticos e Cenários para Adaptação. Relatório/Informe descrevendo o Produto 2 conforme o termo de referencia Contrato 0000019436, CIC Bacia do Prata-SG-OEA.

Su, Z., Cui, M., & Fan, H. (2013). Effect of shelterbelts distribution on ephemeral guly erosion in the roling-hily black soil region of Northeast China. *Research of Soil and Water Conservation, 19*(3), 20–29.

Sun, C., Huang, G., Fan, Y., Zhou, X., Lu, C., & Wang, X. (2019). Drought occurring with hot extremes: Changes under future climate change on Loess Plateau, China. *Earth's Future, 7*(6), 587–604.

Sun, L., Cai, Q., Chen, S., & He, J. (2012). Integrated governing system on soil and water loss of small watersheds in a typical black soil region of Northeast China. *Research of Soil and Water Conservation, 19*(3), 36–41, 57. (in Chinese with English abstract).

Takle, E., & Gutowski, W. J. Jr. (2020). Iowa's agriculture is losing its Goldilocks climate. *Physics Today, 73*(2), 26–33. doi: 10.1063/PT.3.4407.

Terra, J. A., & García Préchac, F. (2001). Siembra directa y rotaciones forrajeras en las lomadas del este: Síntesis 1995–2000. SerieTécnica 125.INIATreinta y Tres, 100 p.

Terra, J. A., García Préchac, F., Salvo, L., & Hernández, J. (2006). Soil use intensity impacts on total and particulate soil organic matter in no-till crop–pasture rotations under direct grazing. In R. Horn, H. Fleige, S. Peth & X. Peng (Eds.), *Advanced Geoecology, 38* (pp. 233–241).

Thaler, E. A., Larsen, I. J., & Qian, Y. (2021). The extent of soil loss across the US Corn Belt. *Proceedings of the National Academy of Sciences, 118*(8), e1922375118. doi: 10.1073/pnas.1922375118.

United States Department of Agriculture. (2020). Summary report: 2017 national resources inventory. Washington, DC: Natural Resources Conservation Service, and Ames, IA: Center for Survey Statistics and Methodology, Iowa State University. Retrieved from https://www.nrcs.usda.gov/wps/portal/nrcs/main/national/technical/nra/nri/results/

USDA Farm Service Agency (FSA). 2021a. Conservation Reserve Program - Fact Sheet June 2021. https://www.fsa.usda.gov/Assets/USDA-FSA-Public/usdafiles/FactSheets/crp-general-signup-56-enrollment%20period-june-2021.pdf

USDA Natural Resources Conservation Service. 2021b. Environmental Quality Incentive Program. Retrieved from https://www.nrcs.usda.gov/wps/portal/nrcs/main/national/programs/financial/eqip/

USGCRP. (2018). *Impacts, risks, and adaptation in the United States: Fourth national climate assessment*, Volume 2. [D. R. Reidmiller et al. (Eds.)]. Washington, DC: U.S. Global Change Research Program. doi: 10.7930/NCA4.2018.

Wang, B., & Zhang, F. (2008). Mechanics and effectiveness of soil and water conservation measures in Northeast China. *Science of Soil and Water Conservation, 1*, 9–11. (in Chinese with English abstract). doi: 10.14123/j.cnki.swcc.2008.01.006.

Wang, J., Chen, C., & Li, T. (2003). Benefit evaluation of watershed management on Loess Plateau of Shanxi Province. *Bulletin of Soil and Water Conservation, 6*, 61–64.

Wang, J. J., Jiang, Z. D., & Xia, Z. L. (2014). Grain-for-green policy and its achievements. In A. Tsuenekawa, G. Liu, N. Yamanaka & S. Du (Eds.), *Restoration and development of the degraded Loess Plateau, China* (pp. 137–147). Berlin: Springer.

Wang, L., Shi, H., Liu, G., Zheng, F., Qin, C., Zhang, X., & Zhang, J. (2019). Comparison of soil erosion between wide and narrow longitudinal ridge tillage in black soil region. *Transactions of the Chinese Society of Agricultural Engineering, 35*(19), 176–182. (in Chinese with English abstract). doi: 10.11975/j.issn.1002-6819.2019.19.021.

Wang, S., Fu, B., Piao, S., Lü, Y., Ciais, P., Feng, X., & Wang, Y. (2016). Reduced sediment transport in the Yellow River due to anthropogenic changes. *Nature Geoscience, 9*(1), 38–41.

Wang, S., & Su, J. (1995). The origin, implement and the others of eco-agricultural development strategy in Baiquan County. *Agricultural Environment and Development, 12*(1), 5–8, 47.

Wang, X. B., Cai, D. X., Hoogmoed, W. B., Oenema, O., & Perdok, U. D. (2007). Developments in conservation tillage in rainfed regions of North China. *Soil and Tillage Research, 93*(2), 239–250. doi: 10.1016/j.still.2006.05.005.

Wei, Y., Li, X., & Hu, T. (2013). Soil and water conservation and water-saving and soybean yield-increasing effects of different conservation tillage technology modes in slopping farmland. *Journal of Northeast Agricultural University, 44*(5), 51–55. (in Chinese with English abstract). doi: 10.19720/j.cnki.issn.1005-9369.2013.05.011.

Williams, C. K., & Rasmussen, C. E. (2006). *Gaussian processes for machine learning.* Cambridge, MA: MIT Press.

Xiaobing, L., Burras, C. L., Kravchenko, Y. S., Durán, A., Huffman, T., Morras, H., Studdert, G., Zhang, X., Cruse, R.M., & Yuan, X. (2012). Overview of mollisols in the world: Distribution, land use and management. *Canadian Journal of Soil Science, 92*(3), 383–402. doi: 10.4141/cjss2010-058.

Zhang, Q., & Li, Y. (2014). Effectiveness assessment of soil conservation measures in reducing soil erosion in Baiquan County of Northeastern China by using 137Cs techniques. *Environmental Science: Processes and Impacts, 16*(6), 1480–1488. doi: 10.1039/C3EM 00521F.

Zhang, Y., Peng, C., Li, W., Tian, L., Zhu, Q., Chen, H. (2016). Multiple afforestation programs accelerate the greenness in the 'three North' region of China from 1982 to 2013. *Ecological Indicators, 61*, 404–412.

Zhang, Z. (2002). Chinese forestry development toward soil and water conservation. In *Proceedings of the 12th international soil conservation conference,* May 26–31, pp. 391–397. Beijing: Tsinghua University Press.

Zhao, G., Zhai, J., Tian, P., Zhang, L., Mu, X., An, Z., & Han, M. (2018). Variations in extreme precipitation on the Loess Plateau using a high-resolution dataset and their linkages with atmospheric circulation indices. *Theoretical and Applied Climatology, 133*(3–4), 1235–1247.

Zheng, H., Liu, W., Zheng, J., Luo, Y., Li, R., Wang, H., & Qi, H. (2018). Effect of long-term tillage on soil aggregates and aggregate-associated carbon in black soil of Northeast China. *PLOS ONE, 13*(6), e0199523.

Zheng, X., Zhu, J., & Xing, Z. (2016). Assessment of the effects of shelterbelts on crop yields at the regional scale in Northeast China. *Agricultural Systems, 143*, 49–60.

Zhou, Y., Li, N., Dong, G., & Wu, W. (2013). Impact assessment of recent climate change on rice yields in the Heilongjiang reclamation Area of north-east China. *Journal of the Science of Food and Agriculture, 93*(11), 2698–2706. doi: 10.1002/jsfa.6087.

Zhu, J., Zheng, X., Wang, G., Wu, B., Liu, S., Yan, C., et al. (2017). Assessment of the world largest afforestation program: Success, failure, and future directions. *bioRxiv* 105619. doi: 10.1101/105619.

Zhu, X. (1991). The formation of the Loess Plateau and its control measures. *Bulletin of Soil and Water Conservation, 11*(1), 1–8, 17.

3 The Psychology of Energy Efficiency

Kristin Riott

CONTENTS

DOI: 10.1201/9781003048701-3

3.1 INTRODUCTION

A friend living in a really old house says, "If you have water in your basement," (pause for dramatic effect), "just don't go down there." I introduce this friend's problem-solving approach as one much favored by *Homo sapiens*: avoidance and procrastination. If the problem is daunting, and especially if it involves getting out the wet vac to suck up backed-up sewage, avoid it. After a couple of months, crack open the basement door and sniff; if you're in luck, the lake has dried up. As problems go, water in the basement has some significant advantages. You can literally close the door on it. And the longer you procrastinate, the more likely that the problem has disappeared.

The basement is the perfect metaphor for our subconscious mind: the place we do not see and would rather not go. And yet go we must, to the subterranean basement or the superterranean attic, should we wish to reduce our environmental impacts, temper the effects of climate change and pocket the savings available from energy efficiency. For American houses and apartment buildings, most of them "stick-built" on site, leak energy like mad. Considering all the cracks, crevices and fissures, which pock the average American house or apartment, it is no wonder that legions of mice laughingly prance in to bask in the warmth. Mice can squeeze through a hole the size of a dime, but air that you paid good money to heat or cool can escape from much smaller exits than that.

3.2 BRIDGING THE GAP OF KANSAS CITY, MO (KCMO)

Before delving further into matters of psychology, energy efficiency, environment and the links between these and climate change, I'd like to establish where my authority, such as it is, lies. First, I am a member of *Homo sapiens*, and therefore, among many other attributes which we will examine in this chapter, prone to procrastination.[1] Second, as a member of the great ape family, I am also prone to mimicry, easily swayed by the behaviors of others in my social group, and attentive to social power and hierarchy. Third, I serve as the leader of a nonprofit organization, named Bridging The Gap (BTG). BTG was established in 1992 in Kansas City and is devoted to environmental education and action through volunteerism.

BTG was founded by Attorney Robert J. Mann, with a vision of building healthy communities and local economies through the power of people working together, especially when drawn from diverse groups such as the business, residential and government sectors. Bob innately grasped that, "Like all social species including humans, it's important for apes to work together in groups to survive, reproduce, and protect themselves from outside threats. The stronger the group, the more likely they are to fulfill these basic evolutionary needs" (Handel, S., 2016).

Very quickly, Bob Mann's emphasis on the facilitation of working together melded with the burgeoning 1990s topic of environmentalism. BTG's very first project was to establish a community recycling center through the use of volunteers, while mobilizing a campaign to bring curbside recycling to Kansas City, MO. Soon, BTG employees and volunteers were traveling by bus through the state of Missouri, educating people about the importance of recycling. Over the next 30 years, BGT absorbed several other nonprofits, while creating new programming. Currently, our programs include recycling center management; business consultation; urban forestry; remnant[2] prairie restoration and management; residential, commercial and industrial energy and water efficiency;[3] business sustainability; and workforce development in green infrastructure maintenance. With so many programs, BTG is able to compare their relative impacts, see how one type of environmental problem intersects with another, and identify the common human attitudes and behaviors which ultimately affect our successes.

BTG has a strong action orientation, doing physical environmental remediation and engaging volunteers. In our work, educating people about and planning for climate change, we advocate whenever we can for physical action, not only because climate change is a physical problem, but because people become ego-invested and confident in solving environmental problems when working with their hands.

All sustainable actions are important and usually offer multiple benefits. Recycling, for example, is job-rich, greatly reduces the energy it takes to produce products from virgin materials and thus reduces the industrial emissions driving climate change, while helping with other problems, like deforestation and the buildup of plastics in oceans. But some environmental solutions have more advantages, cost less and/or are simpler to execute than others. In fact, because there are so many possible courses of environmental action to take, the phrase "bang for the buck" has become a guiding principle at BTG, as a way to prioritize when dollars are limited.

3.3   WHY EVERYBODY SHOULD BE ENGAGED IN ENERGY EFFICIENCY

The perspectives offered in this chapter are informed by BTG's direct experience in conducting public education and installing water and efficiency devices in buildings. We work in, and around, the most challenged urban households, as well as area companies, to prevent wasted energy and water.

Buildings account for about 30% of the emissions driving climate change. It's estimated that 90% of existing U.S. homes are underinsulated, wasting energy and money, while decreasing comfort for homeowners (North American Manufacturers of Insulated Materials, 2015). "If all U.S. homes were fitted with insulation, based on the 2012 International Energy Conservation Code (IECC),[4] residential electricity use nationwide would drop by about 5 percent and natural gas use by more than 10 percent," says Dr. Jonathan Levy, Professor of Environmental Health at Boston University School of Public Health. (U.S. Department of Energy Office of Energy Efficiency and Renewable Energy 2012), Usage of direct combustion fuels in residences, such as propane, would drop even more dramatically (Levy, J. et al., 2016).

McKinsey & Company cites similar numbers, noting that residential savings achieved through more efficient sealing of building envelopes, appliances and electronic devices would account for approximately a 5% decrease from the United States' greenhouse gas emissions levels in 2005 (Granade, H. C., Creyts, J., Derkach, A., et al., 2009). If the commercial and industrial sectors also achieved their energy-efficiency potential, nationwide, the total reduction in greenhouse gases related to energy use would be 14%. While this reduction is obviously a modest proportion of total greenhouse gas emissions, their energy efficiency would be accomplished less expensively and with lower carbon impacts than by conversion to renewables. Additionally, the effort would be multi-beneficial, relatively easy to execute and near at hand.

In fact, energy efficiency in buildings offers enormous potential to meet many other national goals, including our national health. Coal-fired electricity production still occupies a large chunk of Kansas City's (and the nation's) energy capacity. The burning of coal emits not only carbon dioxide, but also a cocktail of free-floating molecules of arsenic, mercury, sulfur dioxide, nitrous oxide and other toxins into the air that we all breathe. Dr. Jonathan Levy, Boston University, specializes in the impacts of air pollution on urban communities. He has demonstrated and quantified that energy conservation has a direct and favorable impact on the health of citizens by reducing air pollutants associated with coal fire-produced electricity, natural gas and direct combustion fuels such as propane. If American homes were insulated to comply with the 2012 International Conservation Code, residents would save US$11 billion directly in reduced residential utility bills, and the value of the associated health benefits from reduced exposure to toxins associated with burning fossil fuels saves another US$2.9 billion, not to mention 320 premature deaths per year (Levy et al., 2016).

Unfortunately, the present state of energy inefficiency in the United States and other countries serves to undermine health. Citizens subjected to chronic cold in underheated homes in the winter months are more vulnerable to other illnesses, while heat exacerbates asthma. In Kansas City, every summer, heat-related fatalities are suffered, sometimes even in homes where an air conditioner was not running because the residents couldn't afford the cost. In July 2012, for example, ten people died of hyperthermia in Jackson County, MO, where KCMO is centered (Mid-America Regional Council, 2015). Energy efficiency can make these homes much more comfortable and protect the health of residents.

3.4   THE RELATIVELY LOW COST OF ENERGY EFFICIENCY

Despite the steadily declining costs of wind and solar power, energy efficiency is still the cheapest way to stop the build up of greenhouse gas emissions driving climate change (Molina, M. 2018). Energy efficiency costs only 3.1 cents per averted kilowatt hour of electricity vs. the next cheapest source at 4 cents per kilowatt hour to use wind energy. Because energy efficiency is cheap and accessible, energy efficiency has been called the "first fuel" in economic development and world energy supply (International Energy Agency, 2016). Improvements in energy efficiency have already played an important role in the nation's energy picture, as physicist and Rocky Mountain Institute founder Amory Lovins notes:

> Few policymakers realize that saved energy is already the world's largest source of energy services, bigger than oil (i.e., 1990–2016 reductions in global energy intensity saved more energy in 2016 than the oil burned in 2016). The public's impression is similarly lopsided. Decreased energy intensity during 1975–2016 saved 30× more cumulative US primary energy than doubled renewable production supplied, yet the ratio of headlines seems roughly the opposite, because renewables are conspicuous but unused energy is invisible.
>
> **(Lovins, A. B., 2018)**

Note the human tendency to give new, exciting and visible solutions, such as wind turbines and solar panels, more attention than invisible, abstract, older ones, such as energy efficiency.

3.5   OUTSTANDING RETURNS ON INVESTMENT FROM ENERGY EFFICIENCY

Some aspects of energy efficiency offer almost astonishingly high returns on investment, or "bang for the buck." Compact fluorescent lamp (CFL) light bulbs are so much more efficient than the old incandescent bulbs that they pay for themselves in as little as nine months. Residential light-emitting diodes (LEDs) bulbs, especially ENERGY STAR rated products, use at least 75% less energy and last 25 times longer than incandescent lighting (U.S. Department of Energy Office of Energy Efficiency and Renewable Energy, 2021). In 2013, BTG found that kits containing water efficiency devices, such as high-efficiency showerheads and faucet aerators, would pay for themselves within three weeks. An US$11 kit installed by the homeowner saves a busy household at least US$15 on their next water bill, according to our anecdotal evidence.

For the residential sector alone, the McKinsey study projects that for a 12-year period (2008–2012), a present value of US$229 billion in up-front energy efficiency investment costs would (have) yield(ed) a present value of US$395 billion in savings. By 2020, energy use in the residential sector would (have been) reduce by 28% relative to a "business as usual" benchmark. The savings and investment costs translate into an internal return rate of return (IRR) on the investment of over 19%. The McKinsey report further identifies the most important residential energy-saving investments, which include sealing ducts, insulating basements/attics, upgrading heating equipment and adding programmable thermostats. These investments deliver the highest IRRs with the exception of the upgrading of heating equipment (Granade, H. C., Creyts, J., Derach, A., et al., 2009). Despite these compelling benefits and though granted, we now have more households and lower use per home, U.S. energy use in the residential sector is still as high as it was in the 1990s (U.S. Energy Information Administration, 2019).

3.6   ENERGY EFFICIENCY AND RATIONAL BEHAVIOR GAPS

The field of behavioral economics applies economic theories to human psychology as a way to explain and predict behaviors. According to this theory, everything we do is a response to a cost-benefit ratio. We make choices based on benefits outweighing costs, or vice versa. If indeed humans behave rationally, as behavioral economics assume we will, why would we not more avidly pursue energy efficiency, with its many benefits and low costs?

Psychologist and behavioral economist Dr. Dan Ariely, however, subscribes to the view that humans are irrational first. As the author of *Predictably Irrational: The Hidden Forces That Shape Our Decisions*, Dr. Ariely states: "[when you] see how much misery there is in the world, the world isn't like this as an outcome of the decisions of eight billion rational people. It's like this as the outcome of eight billion irrational people." He cites texting and driving as one of many examples of our irrationality. Dr. Ariely finds optimism in the idea of looking at our irrationality squarely and building programs that are designed to accommodate it (Ariely, 2008).

3.7   THE DRAGONS OF INACTION

Dr. Robert Gifford from University of Victoria and his colleagues at the Pacific Institute for Climate Solutions in British Columbia have devoted their careers to the science of human behavior, as it relates to environmental progress and climate change. They specifically have studied the question of how the entire world can be threatened by such a massive and destructive problem as climate change, and done relatively little about it for so long.

In his article "The dragons of inaction: psychological barriers that limit climate change mitigation and adaptation," Dr. Gifford proposes a framework to describe why humans are not taking action to prevent or ameliorate climate change. Energy inefficiency is a similarly complex and abstract problem to climate change. It is a problem-within-the-problem, and like climate change, it's everywhere, and its sources are highly diffuse.

Dr. Gifford postulates seven "dragons" of inaction with regard to climate change: (1) "limited cognition," (2) "perceived risks," (3) "ideologies," (4) "discredence (the tendency to dismiss and deny serious problems—one whopper of a dragon)," (5) "sunk costs," (6) "comparison with others" and (7) "limited behaviors" (Gifford, R. 2011). Though not all of his "dragons" apply to energy efficiency, several do, and this may help to explain the energy efficiency gap.

3.8 IGNORANCE, LACK OF AWARENESS AND LIMITED THINKING ABOUT THE PROBLEM

The first of Dr. Gifford's dragons is "limited cognition." As he admits with refreshing frankness, "the human brain has not evolved much in 1,000 years" (Gifford, R. 2011). Humans are awesome at pumping adrenaline and running away from highly visible problems, but not so great at thinking about abstract and non-urgent problems, like energy inefficiency and climate change.

Baby dragons lurk under the seven parent dragons' wings. A baby dragon for limited cognition is ignorance. We could ascribe some portion of our lack of action on energy efficiency to ignorance, or lack of awareness of the potential for energy efficiency to improve our lives. The U.S. Departments of both Energy and Education are concerned enough about our ignorance on the topic to promote October as Energy Awareness Month.

Our ignorance is not aided by the fact that the topic of energy efficiency is complicated, difficult and arcane. According to the economist Ashok Bardhan and his colleagues at UC Berkeley,

[Home energy use] can be remarkably complex and opaque to most property owners …; significant energy use arises in at least three different home systems: (1) heating, ventilation, and air conditioning (HVAC), (2) sealing and insulation, and (3) electric appliances, about which the homeowner has little knowledge. (He or she) will thus require expert advice … to carry out energy-saving investments that are both technologically and financially efficient.

(Bardhan, A. et al., 2013)

Brandon Hofmeister, Wayne State University Law School, names another form of limited cognition, "Human beings have only a certain amount of cognitive bandwidth, and energy efficiency is all too often simply not on our radar screens" (Hofmeister, B., 2010). Work, kids, bills and recreation are priorities, not invisible energy losses in remote, dusty places.

3.9 COMPARISONS WITH OTHERS

One of the highly touted interventions in the world of energy efficiency has been utilities informing their customers how their energy use compares with homes nearby. Not only do we want to emulate others in our social group, but the competitor within demands to know where we are in a social hierarchy. However, Dr. P. Wesley Schulz, Professor of Psychology at California State University, and his colleagues report this comparison strategy can backfire. When homeowners were told the amount of energy that average members of their community used, they tended to alter their use of energy to fit the norm, i.e., decreasing or increasing their energy use accordingly, which is a typical mimicry behavior. Fortunately, the researchers learned that the increases could be prevented by giving low-energy users positive feedback about using less energy (Schulz, W. P., et al. 2007).

3.10   LIMITED BEHAVIORS

About this dragon, Dr. Gifford states, "Some climate-related behaviors are easier to adopt than others but have little or no impact on greenhouse gas emissions" (Gifford, 2011). However, their ease of adoption means these actions tend to be chosen over higher cost but more effective actions. The tendency has also been called the "low-cost hypothesis" (or "tokenism" or the "single-action bias"). The Earth Institute at Columbia University also noted this tendency. "Many Americans believe they can save energy with small behavior changes that actually achieve very little, and severely underestimate the major effects of switching to efficient, currently available technologies" (The Earth Institute at Columbia University, 2010). People typically are willing to take one or two actions to address a perceived problem, but after that they start to believe they have done all they can. At BTG, we've noticed that recycling is not necessarily the "gateway" behavior leading to greater environmental action which it is sometimes supposed to be.

3.11   RISK AVERSION

The most relevant risk related to energy efficiency presented by Dr. Gifford is financial risk and the fear of spending and/or wasting money. Some aspects of energy efficiency, especially when conducted by top-flight professionals, are intimidatingly expensive and require large up-front outlays, such as replacing an old air conditioner or an old furnace. Added to unaffordability is the fact that human minds are essentially risk- or loss-averse. Humans have been shown to value losses at approximately twice the rate of gains (Kahnerman, D., & Tversky, A., 1992). The outlay of dollars spent to replace inefficient heating and cooling equipment, or insulation of a home, may be viewed attitudinally as a kind of loss or sunken costs. It's simply hard for people to imagine that they will get their money back, especially when it can take up to five years for the return.

   In 2015, BTG pulled together a cohort of companies to work collectively in a "Sustainability Circle" that was created by a California-based consulting company. Even though the consulting company's track record clearly demonstrated that dozens of companies profited an average of US$30,000 per year, by saving energy and material, it was very difficult to get each company to contribute US$5,000–10,000 to join the group. The local electrical utility ended up generously subsidizing about half of the companies' cost of participation.

3.12   BEHAVIORAL MOMENTUM OR INERTIA

Dr. Gifford cites behavioral momentum, in particular the "habit of non-action," as one of his dragons of inaction. The tendency of an object to remain at rest, or remain in motion, is known by physicists as "inertia." In terms of human behavior, inertia is "the tendency to keep doing what you're doing," says neuroscientist and behaviorist Dr. John Burkhardt of Columbia University, while adopting new behaviors is more difficult (Burkhardt, J., 2019). Resistance to change is found not only in individuals, but groups of people, as "social inertia." In the creation of a kind of marketplace inertia, buyers tend to hold on to products which no longer are optimal for them or our society. For example, even though the typewriter is outdated now, we still use its famously inefficient QWERTY keyboard design on our computers.

"Behavior seems to follow the status quo unless it is acted upon by a decrease in friction or an increase in fuel," says behavioral scientist, Aline Holzwarth (Holzwarth, A. 2019). To introduce new behaviors and get widespread work on energy efficiency, then, we will have to apply a lot of "fuel."

3.13 PROCRASTINATION

Reinforcing the status quo is the human tendency to procrastinate, which Eric Jaffe of the Association for Psychological Science defines as "the voluntary delay of some important task that we intend to do, despite knowing that we'll suffer as a result" (Jaffe, E., 2013).

Looked at from the standpoint of behavioral economics, we procrastinate when our self-control and motivation to act are outweighed by demotivating factors, such as fatigue, far-off or abstract rewards, anxiety and tasks that are unpleasant. The costs of present actions are outweighed by the attractions and benefits of delay. Paradoxically though, the anxiety and fear of failure which underlie procrastination end up producing even more of the same vicious cycle. The greater the dread of doing something, the longer we wait.

3.14 WHY PEOPLE PROCRASTINATE AND WHAT TO DO ABOUT IT

Dr. David Burns, a retired psychiatrist and professor of clinical psychology at Stanford, practiced in the field of cognitive behavioral therapy (CBT) and hypothesizes that our moods are created by our thoughts, which can be redirected toward more positive outcomes. This is done by examining one's thinking to identify and correct "cognitive distortions" (unhelpful thoughts, beliefs and attitudes) and behaviors. In Chapter 9 of his bestseller *The Feeling Good Handbook*, Dr. Burns asks,

> which comes first—motivation or productive action? If you said, "motivation," ... that's the way a lot of procrastinators think. ... Procrastinators tell themselves, "I don't feel like it. I'll wait until I'm in the mood." Procrastinators comfort themselves in the present with the false belief that they'll be more emotionally equipped to handle a task in the future.
>
> **(Burns, D. D., 1989)**

Dr. Burns puts the lie to this belief, "You're never going to feel like it! These are boring, unpleasant tasks!" Dr. Burns offers an insight that initially seems counterintuitive: "motivation doesn't come first—productive action does. The more you do, the more you'll feel like doing, but doing something comes first" (Burns, D. D., 1989).

Dr. Burns then outlines "A Prescription for Procrastinators":

1. Choose an action to begin to establish momentum and break the tension of inertia.
2. Develop a detailed, specific plan, and identify likely barriers to its completion.
3. Make the job easy. Set modest goals and break into small, doable pieces.
4. Think and speak positively of accomplishments and give credit where it's due. Identify and address negative thinking (Burns, 1989).

Humoring Dr. Burns, let's choose an energy efficiency action for Step 1: gooping up basement ductwork with mastic. The 2009 McKinsey study showed that sealing ductwork is the action with the greatest potential of any intervention for residential energy efficiency (Granade et al., 2009). This keeps the heated air to go up into the parts of the house, keeping the occupants warm and toasty. In Step 2, plan development, vagueness is the enemy and specificity galvanizing. If you can't articulate when you plan to do something, in some sense it's a statement that you don't plan to do it at all. One should write down the exact time to get started gumming things up with mastic. For Step 3, breaking this task down into smaller pieces, the mastic should probably be purchased in advance, slaying at least one baby dragon of inaction.

Dr. Burns also asks us to think in advance of any distractions or problems which might prevent follow-through on our project. This is key, because barriers and hurdles have stopped even a well-intentioned non-procrastinator, noting that their existence often incites procrastination in the first place. We must identify and dispel negative thoughts about icky stuff on our hands, treacherous ladders, boredom and tedium and the like. Instead, focus on the positive: it could really save money, the house will be more comfortable, healthy and affordable and we will have done a little bit to reduce my carbon footprint. But so far, I haven't done it.

3.15 COMPLEXITY, MULTIPLE BARRIERS TO COMPLETION AND MULTIPLE OPPORTUNITIES TO DROP OUT

Ashok Bardhan and his colleagues write, "The complexity of the decision to invest in energy efficiency measures is itself a deterrent … the very process of considering the alternatives and committing to a specific action may be unpleasant and create disutility" (Bardhan et al., 2013). Energy efficiency in buildings is a multitask proposition.

For example, "large air leaks" is sometimes listed as the biggest single category for loss of conditioned air. This category includes attic fans, plumbing conduits within the walls, canned lighting in ceilings, laundry chutes, joints in ductwork, missing door sweeps, chimneys and more. Adding to the barriers, each of these leaks requires its own material and application techniques. Mastic can be used to seal ductwork; clay rope or caulk are used to tighten windows; and metal door sweeps are nailed into doorways. Who knew there was such a thing an attic fan cover or a chimney balloon? The list goes on, and one's drive toward completion may fade a bit with each step undertaken.

Projects requiring large up-front investments may mean some homeowners and building owners will have to borrow money. Going through the processing of taking out a loan itself is an arduous, multistep process, not to mention the monthly payments. All this can become an additional disincentive to energy efficiency.

However, accomplishing energy efficiency by insulating a home is a one-time job. The benefits of insulating go on benefiting for up to 30 years or more. And the entire community benefits from improved local air quality.

3.16 THE SPECIAL CHALLENGES OF LOWER-INCOME HOUSEHOLDS

In the present-day U.S. national crisis in affordable housing, utility bills can make the difference between staying in a home and facing eviction. Kansas City's average income is below the rest of the state of Missouri. Missouri ranks close to 30th state in the nation in median income (U.S. Census Bureau, 2019). Low income, due to the profound, aggregated injustices of centuries, is strongly associated with being Black or Hispanic. Recently, the ACEEE cited utility bill burdens on African-American households in Kansas City (as a percent of household income) to rank 4th among ten U.S. cities with the highest energy efficiency burdens. In their 2016 study, the utility bill burden for Black Kansas Cities was 7.9% of household income vs. 4.5% for the rest of the Kansas Citians (noting that the national average is just over 3%). Astonishingly, the top quartile of the population, in terms of utility bill burdens, was 16.2% for African Americans and 12% for Hispanics (Drehbol, A., and Ross, L, 2016).

There are two reasons why lower income and underserved communities have a higher utility bill burden. Their base incomes by definition are low, and thus their utility bill burden has a higher percentage cost, compared with homeowners with average incomes. But another fact jumps out when studying lower income homes:

> More than one-third of their excess energy burden was caused by inefficient housing stock. Bringing their homes up to median efficiency would lower their energy burden from 7.2% to 5.9%. For African-American and Latino households, 42% and 68% of the excess energy burden, respectively, was due to inefficient homes. For renters that number was 97%, meaning that almost all of their excess energy burden could be eliminated by making their homes as efficient as the median.
> **(American Council for an Energy Efficient Economy (ACEEE), 2016)**

There are many contributing factors which drive up the utility bill burden borne by lower income people. Working in Kansas City, Missouri's lowest-income neighborhoods and homes, BTG has witnessed that the housing stock is older and can tend to have a lot of leaks. Not only is inadequate insulation driving conditioned air loss, but aged plumbing and damaging water leaks are driving up water bills. Additionally, a primary source of energy loss is from holes in roofs which are prohibitively expensive to replace. Holes not only lose conditioned air; they introduce a cascade of problems that include excess moisture, wood rot, mold and potentially mold-driven illness. Aging appliances and older heating and cooling systems also contribute to inefficiency.

Lower-income people and those subjected to systemic racism are not only unable to afford replacing expensive items outright, but are often unable to qualify for loans, or prefer not to. In 2013–2014, the City of KCMO had loan funding available for citizens to implement energy efficiency projects, but there was very low uptake of these funds. A lack of trust borne of harsh experiences with institutions keeps these citizens from participating in energy efficiency programs. In addition, other issues such as lack of transportation, telephones, health problems and physical disabilities create barriers. Finally, it may be more difficult for many low-income people to gather information. In some cases, they may lack computer access or skills, or be unable to read.

BTG has found that safety problems in lower income homes can add further barriers to energy efficiency. In one household, we found carbon monoxide levels in the kitchen that were five times the federally defined safety level, emitted from an old stove. The draftiness of the house was the only reason the occupants were still alive. BTG was able to be a donor of a replacement stove. Problems such as out-of-date, faulty wiring in older and less well-maintained homes can greatly impede progress by adding more steps, and more layers of cost and complexity.

Even where there is the knowledge and the will to make energy efficiency improvements, lower income households may find that they can't afford energy-efficient items such as new LED light bulbs. As James Baldwin wrote, "Anyone who has ever struggled with poverty knows how extremely expensive it is to be poor" (Baldwin, J., 1961).

3.17   LOWER INCOMES AND ENERGY INEFFICIENCY IN APARTMENT BUILDINGS

About 36% of the U.S. population live in rental properties. Within this sector, 37% live in multifamily housing, 40% in houses and 23% in mobile homes, duplexes or quadraplexes (National Multi-Family Housing Council, 2021). Though their square footage is often much lower, the utility bill burden borne by those renting their homes, as a percentage of disposable income, is heavier than among average homeowners. The national average income of renters overall is much lower than in owned homes, at US$39,000 per year median income vs. US$73,000 for the average homeowner (Harvard Joint Center for Housing Studies, 2017). Nearly half of the nation's population of renters is composed of people of color, twice the incidence found in home ownership demographics. Furthermore, apartment renters tend to be younger and have substantially lower net worth than homeowners. Therefore, energy efficiency investments of any kind would be more difficult for these renters to undertake financially and perhaps psychologically as well.

As with lower-income homeowners, another reason utility bill burdens are so high among renters is the inefficient state of their housing:

> most (but not all) MFBs[5] are/were designed to be built with the lowest possible first cost. They then operate in a chronically undercapitalized environment. As a result,

these buildings have energy-wasting shells and systems, and rarely see energy improvements as they age.

(Hynek, D., Levy, M., Smith, B. et al., 2012)

Split incentives can potentially be a negative improvement when landlords are not being motivated to improve energy efficiency when renters are paying the bills, and thus can become a powerful barrier to energy efficiency in apartment buildings.

3.18   ENERGY EFFICIENCY IN RURAL AMERICA

In rural America as elsewhere, low incomes are strongly associated with energy-inefficient housing. "Much of rural America has still not recovered from the great recession" (National Rural Housing Coalition, 2021). Poverty levels in rural America are 4.5 percentage points higher than in metropolitan areas (U.S. Department of Agriculture Economic Research Service, 2021). Housing stock in most rural areas today is generally older (Ross et al., 2018), built before small town populations dwindled in the 1950s, when building codes were more lenient and thus less efficient. Rural households have a median energy burden of 4.4%, compared to the national burden of 3.3%. But low-income rural households experience the highest median energy burden at 9%, almost three times greater than the national average. In several regions, one-quarter of the low-income rural households have a median energy burden greater than 15%. Elderly, non-white and renting households face particularly high burdens (National Rural Housing Coalition, 2021).

To summarize about low-income households, barriers which prevent them from engaging in energy efficiency projects are cost, decision-making complexity, the remoteness and uncertainty about payback, the lack of full information and the multistep nature of the work. These barriers are exacerbated in communities that do not have access to technology, and therefore not able to gain education and information. Poverty itself (not the stress of poverty, which is yet another factor) has been shown to be very mentally and physically exhausting (Kelly, M. 2013), diminishing the capacity to tackle complex projects.

3.19   POLITICAL AFFILIATION

Unfortunately, but understandably, there is at least one more reason why some people might not concern themselves with energy efficiency. Environmentalism is associated with the left end of the political spectrum. BTG has found that some people are unwilling to recycle, for example, because they might be advancing a liberal cause. Climate change, in particular, has been politicized since Al Gore was so strongly associated with it in the early years of public awareness. As a result, there has been a most unfortunate delay in mobilizing our country to deal with climate change.

3.20   EXISTING ENERGY EFFICIENCY PROGRAMS
AND THEIR RELATIVE SUCCESSES

Utility companies, particularly the electrical power companies, are motivated to avert the heavy cost of building additional power plants. In some cases, state laws require them to engage customers in energy efficiency initiatives. As a result, electrical companies have introduced many programs and in multiple states, encouraging the adoption of energy efficiency measures. The electric utility company in Kansas City has spent several years installing "energy-saving kits" in homes. The kits contain efficient light bulbs, water heater insulation, "smart strips" that facilitate turning off multiple appliances at once and water efficiency devices, such as low-flow showerheads. Low-flow showerheads reduce both the use of electrical heating of water in homes and municipal electricity use to pump water into homes. Because of long working hours, frequent moves, phone shutoffs and other factors, electric companies across the country have found it difficult to communicate with residents of lower income neighborhoods. In Kansas City, for example, the zip code area of 64128 has a median income of US$26,535 and contains more than 6,000 households (Income by Zip Code, Inc., 2021). Unfortunately, only 135 households received the energy-saving kits in the current year.

City governments have become quite active in passing energy efficiency legislation and ordinances, as awareness of the threat of climate change has increased. In 2016, Kansas City passed an ordinance requiring owners of buildings over 50,000 square feet to benchmark their building's energy efficiency, using the U.S. EPA's and the U.S. DOE 's ENERGY STAR Portfolio Manager Program (U.S. Department of Energy & U.S. EPA, 2021). The idea was that once building owners were graded on a curve with the other building owners and learned how inefficient their building(s) were, they would become motivated to take action and insulate their buildings. But so far, only about half of the building owners have complied with the ordinance, and very few have gone on to improve their energy efficiency. Stronger enforcement measures are being developed.

U.S. EPA/U.S. DOE's Portfolio Manager requires hours of work in the creation of energy benchmarks involving multiple buildings, and the benchmarking needs to be repeated every year or so to gain insight into the buildings' energy consumption. Once these activities are completed, the psychological bandwidth to go on and complete energy-saving projects may be somewhat depleted. BTG, in tandem with an engineering consulting company and with support from the local electric power company, works to encourage these businesses to keep going.

In many places across the country, cities and utility companies join their energy efficiency efforts with nonprofit groups like BTG. Various combinations of these entities have experimented with programs which focus on behavior change, such as asking people to turn the lights off when they leave the room and turn off computer(s) overnight. One of the most common ways to run a volunteer-based program is to introduce a contest focusing on energy conservation, known as "gamification." The idea is to take a relatively boring activity and make it fun and social. The ACEEE published a summary of 53 such games from across the country. In 2010, the Climate and Energy Project (a nonprofit organization near Kansas City), used gamification to compare the energy use of six Kansas towns against six "control" towns. The reduced use of energy was supported using public education and local media. Overall, the project produced an average reduction in energy use of 5.5%, a typical result. There were ancillary benefits, such as the school superintendent noticing a huge difference between the energy use from one school to another. Smaller, tight-knit groups might be able to achieve a larger reduction goal, such as 10%, according to the assessment of all 53 games (Grossberg, F., Wolfson, M., Mazur-Stommen S., et al. 2015).

One question about gamification is whether the newly habits introduced are lasting. Evidence shows that the behaviors persist for two years, then "backsliding" occurs until at five years, the habits have largely dissipated (Sussman, R., Gifford, R., & Abrahmse, W., 2016). Considering the costs of gamification, which in one example was approximately US$175,000 per year (Grossberg, F., Wolfson, M., Mazur-Stommen S., et al. 2015), as well as the complexities, more permanent solutions that don't require behavior change, such as vacancy sensors in classrooms and insulation in apartment complexes, could be given priority.

Gamification can create other benefits beyond energy savings. Chief among the outputs are community relationship building and social cohesion, while helping people understand the many factors which affect the ways in which environmental damage can be prevented and how environmental progress can be made.

3.21   MUNICIPALITIES AS CATALYSTS IN RESIDENTIAL AND BUSINESS ENERGY EFFICIENCY

Energy efficiency programs would result in heightened comfort, health and lower energy bills for some of our most challenged citizens, and help to avoid eviction, providing a compelling argument for local governments to introduce such programs quickly. Energy efficiency offers the potential to contribute to the stability of communities and to right the long-term injustices in communities of color and lower-income areas.

Encouraging energy efficiency in the private sector is not common in city programming; city governments have their hands full taking care of public lands, streets and properties. The internal culture of many cities have evolved over time as a response to whatever citizens complain about the most. "Complaint-based" systems abound in municipal governments identifying a range of problems such as potholes, illegal dumping, burned-out streetlights, graffiti and fallen tree branches.

Complaint-based systems, however, do not always serve citizens in an equitable manner. People in vulnerable communities are typically working long hours, may have difficulty making calls to city government during working hours or have to navigate too many other bureaucratic systems to bother with a complaint. Most importantly, citizens can't complain about what they don't know much about, such as energy efficiency. But without intercession by governmental entities, lower income citizens are consigned to endless physical discomfort, subjected to safety issues and deterioration of the physical intactness and value of their homes from rain, mold and decay. From BTG's experience, this is how residents with incomes of US$25,000 or less presently live in the United States. In Kansas City, that is approximately 22% (City of Kansas City, Missouri, 2019).

Enactment of policy change, such as the implementation of robust building codes, is one way that municipalities can affect the nation's progress with energy efficiency, but this doesn't address the vast majority of existing homes which are underinsulated. Municipalities have the institutional credibility and the ability to communicate with their citizens and can and should play an important role in popularizing energy efficiency.

3.22   HOW CHANGE HAPPENS: HELP FROM A WARM, KNOWLEDGEABLE FRIEND

One way to get something difficult done has always been to ask a friend over for moral support. If my kind and friendly neighbor would come over and help, we might get that mastic gooped on after all.

Dr. Gawande, a MacArthur Fellow and Assistant Professor of Surgery at Harvard Medical School, is also a 2006 recipient of the "Genius Award" from the MacArthur Foundation. In a New Yorker article, "Slow Ideas," Dr. Gawande tells the story of two breakthroughs in medical science that happened at about the same time in the mid-1800s (Gawande, A., 2013). The first was the introduction of anesthesia, and the second was antisepsis, or sterilization of instruments, surfaces and doctors' hands, which greatly reduced the risk of infection. Within just a few months, anesthesia was being used by physicians around the world. But 170 years later, antisepsis and sterilization have still not been fully adopted.

Dr. Gawande offers ideas as to why the rates of adoption of these two medical breakthroughs should be so radically different. The use of anesthesia was ripe for rapid adoption. Ether was inexpensive, and fairly easy to administer. Surgery could be conducted without the visible and audible agony and movements of patients, a major and direct benefit to the doctor. Antisepsis and sterilization, on the other hand, required painstaking attention to detail. The rewards were not immediate and visceral, but rather offered the abstract potential for warding off further sickness. To this day, sterilization has not been fully adopted in countries like India; not even handwashing is regularly practiced among caregivers in rural hospitals. Dr. Gawande states, "Many important but stalled ideas ... attack problems that are big but, to most people, invisible, and making them work can be tedious, if not painful" (Gawande, A., 2013). Climate change is such an invisible and challenging problem; energy efficiency is another.

The solution to tackle complex problems advocated by Dr. Gawande is simple: introduce a warm, non-threatening, but knowledgeable person to assist in overcoming barriers and adopting new norms. "People follow the lead of other people they know and trust," says Dr. Gawande. As we have noted, the human tendency toward mimicry is useful here as well. The introduction of a teaching nurse in rural Indian hospitals showed that sterilization would be adopted in as little as two weeks. A warm, non-threatening manner and good listening skills improved outcomes.

In the world of climate change solutions, an energy auditor is probably the closest thing we have to a warm, knowledgeable friend to come in and break our inertia. This is the person who comes to your house and uses infrared cameras or seals up your front door with a giant blower, to reveal where air is leaking. BTG has debated the use of audits, in part because they are expensive (up to US$1,000, though sometimes utilities offer free ones), money which could be applied toward sealing rim joists and insulating attics, which every house needs anyway. The Kansas City Metropolitan Energy Center,[6] however, has argued that a home should always have an audit, because buildings are like individual organisms, idiosyncratic in the way their air flows. The Pacific Institute for Climate Solutions has this to say about audits:

> The programs earn large-scale savings if they successfully persuade customers to invest in additional major energy-saving products. To this end, energy auditors that do more than simply provide information about which products and rebates are most effective; their personal attention is also vital. A review of energy advisor (auditor) programs concluded that advisors should guide customers through three types of barriers: information barriers (providing knowledge about actions that can save energy and associated rebate programs), decision-making barriers (e.g., reviewing results with the customer), and transactional barriers (e.g., scheduling and paperwork).
>
> **(Sussman, R., Gifford, R., & Abrahmse, W., 2016)**

The personal attention given to the homeowner during an audit may increase the adoption of energy-efficient efforts, making the investment worthwhile.

3.23 DESIGNING AN OPTIMAL ENERGY EFFICIENCY PROGRAM: HIGHEST IMPACT MEASURES AND INCORPORATING FRIENDS

About 70% of U.S. residential energy goes for water heating, space heating and cooling (U.S. Energy Information Administration, 2015). There are several major ways by which heating and cooling efficiency can be achieved: installing new, more efficient heating and cooling equipment, using programmable thermostats to heat and cool when people are home and awake and insulating/tightening the building envelope.

The McKinsey study showed that thermostats, basement insulation and ductwork sealing offer both more potential and at lower cost than maintaining furnaces, heat pumps or air conditioners (Grande et al., 2009). Since we're interested in bang for the buck, let's skip over furnaces and air conditioners for now and deal with the other three. Programmable thermostat installation is one of the highest-impact, lowest-dollar steps that can be taken. As an additional incentive, some

> utility companies give away thermostats for free. However, only 12% of the nation's 118 million households had a central air-conditioning unit that is actually controlled using the programmed thermostat. About one in three households using central air conditioning do not have a programmable thermostat. But even for those households that use central air conditioning and have a programmable thermostat, more than two-thirds of those households control temperatures without actually programming the thermostat.
>
> **(U.S. Energy Information Administration, 2017)**

In other words, people don't know how to program their thermostat, or simply don't bother to do it.

The relatively recent invention of "smart" (self-programming) thermostats will help make programmable thermostat use more widely adopted. Personal temperature preferences and schedules are chosen when the thermostats are installed, so the homeowner need not make any adjustments. The use of "smart" thermostats is especially important in apartment buildings, where insulation opportunities are limited to the landlord. Residents can save several hundred dollars a year, a very meaningful amount in lower income households, with almost no effort.

3.24 LARGE AIR LEAKS

In our continued pursuit of "bang for the buck," while designing our "Housewarmings" program in 2013, BTG focused on sill plate/rim joist areas, located in the basements of homes where the wooden frame of the house rests on the concrete foundation.[7] The 2009 McKinsey study shows such basement insulation right after programmable thermostats as one of the best home energy efficiency investments (Grande, H.C., Creyts, J., Derkach, A., et al., 2009). While the attic floor is an important place to insulate because heated air rises in winter, we decided to insulate rim joists instead. In older attics, electrical wiring often must be updated before insulation can be added, which can be expensive. Also, working in the attic can be dangerous. Even seasoned insulation professionals slip and fall through the rafters and sometimes through the ceiling of the house. We therefore recommend rim joists as one of the first places to look for major heat losses, along with the mastic gooping of ductwork. In 40 Kansas City lower-income homes, Housewarmings interventions reduced the heating bills by an average of 23%. None of the basements in the program were finished or insulated. If they had been, insulating foam could have been sprayed into holes drilled in drywall to increase efficiency.

There are many other sources of large air leaks, especially in older homes. Ill-fitting windows, or single-paned windows, are other common source of leakages. However, storm windows can be added, and casements can be caulked, as opposed to going to the full expense of replacing windows.

As a very important note, it is always necessary to test for carbon monoxide and/or radon before insulating. The health of residents can be negatively affected by sealing a home up too tightly, if carbon monoxide and radon are present or if there are smokers in the household. According to Dr. Lindsay Underhill at Boston University, ventilation levels should be evaluated when extensive insulation is undertaken. Please refer to her work on optimizing insulation vs. ventilation in lower income apartment buildings in particular (Underhill, 2018 and Underhill et al., 2020).

3.25 A NOTE ABOUT MISCELLANEOUS HOUSEHOLD DEVICES

Miscellaneous electronic devices are the most rapidly growing category of household energy use (Bailey, A., 2016). The 2009 McKinsey report identifies efficient miscellaneous electronic devices as televisions, computers, mobile phones, toasters, coffee makers, baby monitors, home security systems and more. The average U.S. household utility bill[8] in 2015 was approximately US$1,500 per year, with a whopping US$420, or 28%, spent to power these miscellaneous electronic products (Fanara, A., Clark, R.,

Duff, R., & Polad, M., 2006). The energy efficiency potential for miscellaneous electronic devices will need to be largely addressed by appliance efficiency standards. There are thousands of electronic products and they become obsolete fairly quickly. Incentives may be introduced to turn in less efficient appliances and/or dispose of them responsibly. Education aimed at electronic device bill relief for lower-income households is also needed, perhaps along with gamification programs.

3.26 TREES NEAR BUILDINGS FOR ENERGY EFFICIENCY

Trees have a powerful effect on energy efficiency, with a host of ancillary benefits. A healthy tree canopy can cool a whole city by 7 degrees or more on a hot summer day and cut the energy use of an individual home by 15–50% (Pandit, R. & LaBand, D. N., 2010). No wonder that cities and countries all over the world are planting trees as a way to prepare for rapid warming from climate change.

The nonprofit Heartland Tree Alliance was founded in the wake of a terrible ice storm in Kansas City in 2002. During the storm, what sounded like gunshots turned out to be the cracking of giant tree limbs caused by the weight of the ice. Heartland Tree Alliance set about replacing the thousands of trees that were destroyed, and later became a program of Bridging The Gap, planting tens of thousands of trees. Now, nearly two decades later, we know much more about the enormous infrastructural and health benefits which trees confer to communities, and planting efforts are increasing further in our region.

In 2018, BTG helped lead the development of Kansas City's Urban Forest Master Plan. One of the plan's rallying cries is "a shade tree for every building." For optimal reduction in heat-related energy use, large, deciduous shade trees should be placed about 10–20′ from the house, to shade east-facing walls and windows from 7 to 11 a.m. and west-facing from 3 to 7 p.m. during June, July and August (Kuhns and Miller, 2011). In general, evergreen trees placed on the northeast and north sides of a house can serve as windbreaks, reducing the utility burden of heating a house in winter. (Wind patterns differ, depending on local topography and other factors.) Such properly placed trees can begin to reduce utility bill burdens within just a few years of planting. In homes where people don't have an air conditioner, or can't afford to run one, a shade tree can literally make the difference between life and death.

Trees have very positive psychological effects on people. It has been shown that having a tree outside the window can speed up healing in a hospital setting. Trees near the workplace can reduce sick days at work, students concentrate better when studying near trees and tree-lined roads reduce road rage incidents. Shade trees can increase real estate values and improve air quality near homes, as well as reducing utility bills (for all tree benefits list above, see Turner-Skoff and Cavender, 2019). The economic value of trees far outweighs their planting and maintenance costs, returning US$2–5 for every dollar spent (U.S. Department of Agriculture Forest Service, Pacific Southwest Research Station, 2011), and should be a public priority to capture all of these benefits.

There are psychological barriers to planting trees, such as a desire not to rake leaves, or worries that pruning and trimming will be too expensive. Public education is much needed to understand trees' many benefits. A highly visible tree-planting, combined with Energy Awareness activities, could be an excellent annual event in any city or town. In Kansas City, the best time for most trees to be planted is in the fall, when there will be a long establishment period before the heat and drought of the following summer. Fall is also an ideal time to launch energy efficiency programs, before cold weather sets in. October is Energy Awareness month in the United States! A small tree whip could be the reward for every home where energy efficiency improvements have been made. Trees have the highest survival rate and health when they are planted as young and small as possible.

3.27   SETTING UP A MUNICIPAL PROGRAM FOR ENERGY EFFICIENCY

All of the energy-efficient activities that have been mentioned are relatively inexpensive. They don't require diagnostics, and they are discrete activities which can be conducted by municipalities or nonprofits.

At BTG, a crucial step after the rough outlines of a program are designed is the hiring of the right energy efficiency program manager. He or she needs to have strong interpersonal skills and feel comfortable with a host of client personalities. Additionally, the program manager requires a willingness to descend into dirty basements or attics and do the hard work of making them airtight. For Housewarmings, BTG hired an easy-going Vietnam veteran, who is literally everybody's friend. We look also for energy and self-motivation when we hire. Being around energetic people conveys a can–do, positive attitude which is very infectious and is the antidote to procrastination. The program manager must work in collegial fashion with local contractors and be ready to book volunteers and schedule home visits. Off-the-shelf applications, developed for the energy efficiency industry, are available for appointment scheduling. The local candidates can help research the most effective efficiency interventions while noting the need for materials and tools such as foam insulation canisters, wand applicators, foam board and pots of mastic, and the development of a protocol for building visits.

BTG's Housewarmings Project employment costs are approximately US$80,000 for a full-time employee, insurance, materials and other expenses. In our first year, federal dollars covered a similar amount for the participation of auditors and contractors. Our program manager was able to insulate 40 urban core homes in the first year. We expected that in subsequent years, once the program is well established, the number of homes insulated might eventually double, with program expenditures of approximately US$1,000–2,000 per house. Additionally, materials purchased by the city at wholesale prices would probably amount to another US$250. Labor costs could be reduced further by family and friends (which was a requirement for Housewarmings), volunteers or municipal workers who are underutilized, especially during the winter months. Residents could also contribute to cost sharing on an ability-to-pay basis, or in lieu of volunteerism. Furthermore, residents might be willing to donate toward other households where residents are not able to pay. Local nonprofits could be hired to manage the program, the volunteers and the incoming donations.

A big cost of any municipal program is the time and travel involved in getting to and from the homes. It is therefore ideal to work on more than one home in a given neighborhood, on a given day. We count on the neighbors to pass information, via "word of mouth," to help spread energy efficiency interest in the neighborhoods. Publicity could notify neighborhoods in advance that the program manager is coming on a specific date. Residents are encouraged to clean out their basements, attics or other work areas in advance, perhaps with bulky item curbside pickup provided. If engagement of others in energy efficiency can be made highly visible and obvious, the likelihood is that energy efficiency efforts would increase.

Where affordable, and in a community-wide effort, local HVAC and insulation contractors can and should play an important role in energy efficiency initiatives, especially to help supervise and train groups of people in the communities. At BTG, we seek minority-owned contractors as a means to advance equity. The expertise and experience of contractors add a great deal of value, while raising the caliber of the work completed.

3.28 MAKING IT EVEN MORE SOCIAL – VOLUNTEERS CREATE VISUAL BEEHIVES AND BUILD SOCIAL COHESION

Introducing more people, especially volunteers, to any energy efficiency effort creates a visual beehive of activity and increases word-of-mouth communication. At BTG, we think volunteers are the most wonderful, hardworking, unselfish people in the world, and we engage them in most of our programs. Volunteer participation needs to be made easy, relatively short (two hours or so) and administratively painless. We try to show our new recruits a good time so that they'll come back for more. The only prerequisite that is required for most volunteers is that they are friendly, reliable in showing up and willing to work, but skilled volunteers, such as a carpenter who can seal up large air leaks, are extremely valuable.

Volunteerism can have a profound psychological impact on participants. It increases the sense of belonging so important to social species, and also increases the social cohesion of any community, weaving people together and providing them with a sense of belonging, which is an important prerequisite of mental health. Social cohesion has been shown to be a leading factor in the survival of residences in dangerous situations, such as the Chicago Heat Wave of 1995. Quoting Danielle Baussan, Center for American Progress,

> Researchers found that 3 of the 10 Chicago neighborhoods with the lowest rates of heat-related deaths were low-income, African American communities. The reason that communities with similar demographics fared so differently was high levels of community interaction and organization that decreased isolation among residents.
>
> (Bassan, D., 2015)

The fact that residents know each other, in other words, is a great predictor of survival in a disaster situation. Neighbors will check on people in the community to make sure they are okay. In the current climate change era, where natural disasters of all kinds are becoming commonplace, lower income groups and people of color are being disproportionately affected. If the Chicago Heat Wave is any indication, much can be learned from these groups about social cohesion.

In business settings, with BTG's Green Business Network, we have found that establishing a green team within a company is one of the most effective ways to ensure that a company is moving forward with energy efficiency and other forms of sustainability, such as waste reduction and "green" purchasing practices. Green Teams and their work can add social cohesion to corporate settings, adding to the enjoyment and meaning of being at work (not to mention free pizza). Opportunities to communicate with senior leaders and their praise can be especially meaningful for employees.

The U.S. DOE has many programs supporting energy efficiency in different sectors, including the promotion of treasure hunts to find energy efficiency in churches and businesses (U.S. Department of Energy Better Buildings Program, 2021). The U.S. DOE also has a consulting network for small–medium-sized manufacturing companies, called Industrial Assessment Centers, where a university professor and graduate students will evaluate the efficiency of a factory or other industrial setting, at no expense to the company (U.S. DOE, 2021).

3.29   SLAYING THE DRAGONS OF ENERGY EFFICIENCY INACTION

With so many dragons in the road, and so many human tendencies like procrastination, inertia, becoming overwhelmed, being sure you can't afford it and quitting in the middle, achieving our country's energy efficiency potential requires us to pour a lot of "fuel" on the fire. Unless energy efficiency is vaulted to the forefront of public discourse, and pursued in a highly visible and energetic manner, we will miss this substantial opportunity in the reduction and aversion of greenhouse gas emissions.

Undeniably as a nation, we have postponed the implementation of climate-related solutions for several decades now. Using ideas that are presented in this chapter, while throwing in a few BTG principles, the starting point for our climate action efforts would be to gather a cross-sectoral group of climate-concerned community members. The first energy efficiency priority should focus on the most heavily burdened, lower income, multifamily rental properties.

3.30   GET STARTED, ESTABLISHING MOMENTUM
AND A STRONG ACTION ORIENTATION

At this moment, cities around the world are working assiduously on their climate change adaptation and resiliency plans. The greater Kansas City's regional plan was put together in 2020 by a nonprofit called Climate Action KC,[9] along with the Mid-America Regional Council.[10] Universally, such plans are naming energy efficiency work as one of the building blocks for climate change readiness while averting climate's most severe effects. However, the plans do not always identify and conquer the barriers described in this chapter. Using Dr. Burns' advice toward specificity, vagueness can be dispelled by appointing leaders that are responsible for particular plan elements, and finally, setting the dates for completion so that the momentum continues. Additionally, it is important to identify detailed elements for the plan, while identifying possible partners, ranging from utility companies, to nonprofits, to for-profit, real estate companies, bankers, etc.

3.31 HIRE OR NAME PERMANENT CITY STAFF FOR ENERGY EFFICIENCY AND THE DEVELOPMENT OF THE PROGRAM

It is very important to hire an "Energy Efficiency Czar," i.e., Dragon Slayer, who will work to convene an influential steering committee. If a written, detailed document doesn't exist, the Czar will lead the energy efficiency plan's writing. This new plan should have real-life programming, while using "bang for the buck" principles, focusing on the highest-impact energy efficiency measures.

One of our immediate tasks is to establish an influential Energy Efficiency Council of utility company leaders, community or government leaders, landlords, building occupants or neighborhood leaders, energy efficiency engineers or HVAC or insulation tradespeople, and businesspeople such as mortgage bankers. It will also be important to recruit leaders from targeted neighborhoods. Marketing people who can create appealing advertising and/or public education campaigns, and preferably have experience succeeding in lower income neighborhoods and with communities of color, are also essential. All members, including a robust diverse representation, should be selected for their ability to work on the problems in a collegial and constructive fashion. The council should familiarize themselves with local building codes and their energy efficiency requirements for stringency, while considering updating their requirements. It will also be important to determine the landlord's responsibilities that are required by law, and whether the codes have been enforced.

3.32 ESTABLISH FUNDING PLANS

Funding projections and fundraising plans are the most critical part of the project and can be one of the most common barriers to real progress. Funding is one of the main reasons for the establishment of the influential and diverse Energy Efficiency Council. Fundraising requires grant writing, requests for funding from private family foundations, loans and city, state or federal fund applications. Thorough research is necessary to discover the best sources and practices for energy efficiency funding. As an example of innovative funding practices, in Europe, there have been experiments with adding Energy Service Companies (ESCOs) at the neighborhood level. ESCOs make energy efficiency investments with no cash outlay from their clients and are paid back through the energy savings. Their business model typically works at larger scales, such as in school districts or large companies. However, the model also works for residential neighborhoods where homes are banded together and treated as one entity.

If incremental funding is not available for energy efficiency work, it may be possible to refocus existing city staff to create a program and see what can be done with volunteers and donated materials.

3.33   IDENTIFY LOWER INCOME APARTMENT BUILDINGS IN THE COMMUNITY AND CONTACT INFORMATION FOR THE LANDLORDS/OWNERS RESPONSIBLE FOR THEM

Landlords of lower income apartment buildings are guardians at the gate of the most egregious energy efficiency problems in America. Cheaply built apartment complexes are filled with low-income residents who are challenged to make ends meet. However, apartment buildings offer a specific benefit to program managers. By only dealing with one person, such as the landlord or building manager, it is possible to improve the lives of dozens of tenants.

Our nonprofit partners at Elevate Energy in Chicago (an energy management system provider) have demonstrated that the "split incentive" between landlords and tenants may have been overemphasized; landlords, indeed, have a financial interest in energy efficiency. In an article about their efforts, Kari Lyderson writes:

> Weatherization and other investments often mean more desirable housing that translates to lower tenant turnover, a big plus for owners. Landlords also stand to save money from utility bills for common building areas or in cases where they pay for heat or other portions of the utility cost .... And energy efficiency upgrades can also mean lower maintenance costs for landlords, especially in terms of new appliances like more efficient furnaces and hot water heaters.
>
> **(Lyderson, K., 2014)**

Engaging landlords in the Energy Efficiency Council, especially in rental apartments, and educating them about energy efficiency should be a high priority because of their ability to affect so many lives, as well as the utility burdens described earlier. Whatever energy efficiency measures are achieved, negotiations must ensure that the landlords pass at least some of the savings on to the lower-income residents.

3.34   REVIEW BUILDING INSPECTION REPORTS AND/OR AUDIT BUILDINGS

It is time to develop a working plan by addressing the first buildings known to be inefficient, using benchmarking data, or possibly a review of city records of buildings which are not code compliant, or where many complaints have been made about landlord practices. Energy audits of some of these lower income apartment buildings would be a second step, and actions identified to insulate them, place storm windows and sealers around window casements and install programmable thermostats. "Retro-commissioning" of these buildings may be undertaken by an ESCO with no cash outlay. This means reviewing their systems to ensure they are operating as closely as possible to the original design/engineering intent.

At the time of this writing, apartment buildings in New York, Canada and elsewhere are being retrofitted to the Passive House standard, promulgated by the Passive House Institute U.S.. Passive housing has unusually deep wall cavities filled with insulation and a tightly sealed envelope, which can reduce heating and cooling loads by as much as 90%, compared to typical existing buildings (Center for

Ecotechnology, 2021). It is possible to reclad either the exterior or the interior of an apartment building or house to create deeper walls and meet the Passive House standard, which is becoming the new "gold standard" for sustainable housing and should be investigated for retrofit projects.

3.35 DEVELOP AND LAUNCH AN ONGOING, YEAR-ROUND PUBLIC EDUCATION AND VOLUNTEER ENGAGEMENT CAMPAIGN

Creating an ongoing "buzz" about energy efficiency is necessary long-term for public education and action. As one method, internships or volunteer opportunities can be offered to young people willing to do the social media work as a means to create visibility and momentum, even without much funding.

Finding the most effective ways to reach notoriously hard-to-reach, lower-income folks requires experimentation and persistence. Door-to-door canvassing can be effective, because people are often reluctant to answer the phones for numbers they don't recognize. Canvassing is also a good way to engage volunteers. To create credibility, volunteers should wear program insignia and carry leave-behind print materials, such as program endorsements of well-known and trusted organizations. If funds permit, it can be effective to hire people, or recruit volunteers from the targeted neighborhoods, as program ambassadors. Selecting a block captain, who has agreed to coordinate with neighbors for a modest stipend, is another cost-effective possibility. A small giveaway to the tenants, such as a CFL light bulb, can create appreciation for the program.

3.36 MAKE ENERGY EFFICIENCY HIGHLY VISIBLE, PREFERABLY IN AN UNUSUAL WAY

To increase awareness and interest, we need to make energy efficiency and its solutions more visible. There are dozens of studies showing that humans remember information much better when there are strong visual images associated with it. As Robert E. Horn states in *Psychology Today*, "(the) effective use of visuals can decrease learning time, improve comprehension, enhance retrieval, and increase retention" (Kouyoumdjian, H., 2012). An effective commercial about climate change and energy, which I remember well, involved black balloons flying up and out of a house. They represent the smoky carbon molecules, lost by energy inefficiency, building up in our atmosphere. Visuals can increase learning by as much as 400% (Shift eLearning, 2014).

An attention-getting energy efficiency vehicle, emblazoned with a program name and emblems at the very least, should become a familiar sight in neighborhoods where energy efficiency programs are underway.

3.37 ITERATE

One of BTG's energy efficiency leaders pointed out that the willingness for iterative work serves our mission well over the long term. Iteration is an antidote to perfectionism. Therefore, we should always be willing to go further in the homes that we have worked on before, especially when a good working relationship has been previously created.

3.38   WORK ON MULTIPLE FRONTS

Dr. Ariely, the expert on irrationality in humans, points out, "(o)ver the last 30 years, we got people to wear their seatbelts. Why? It was a combination of annoying beeps from the car, heavy fines and kids in the backseat screaming if their parents didn't put their seatbelts on" (Airely, D., 2008). Similarly, energy efficiency and climate change solutions both will need a lot of prodding and incentives to counter inertia – a full-court press of public education, financial incentives and social and cultural support from multiple directions to succeed.

As a nation, we need cognitive behavior therapy, to keep us from procrastinating and not allow our fears about climate change to stop us dead in our tracks. Hopelessness about problems like climate change, energy inefficiency and even poverty is a kind of irrationality. The solutions to these problems are actually within our grasp. We only lack a change in objectives, redirection of dollars and the will to make a behavior change, before change can actually happen. When I have dark nights of the soul about climate change, I think of Winston Churchill in the darkest days of World War II, saying, "never give in, never give in, never, never, never, never" (National Churchill Museum, 2021). We don't know yet whether we can hold off the worst effects of climate change, but we are going to go down fighting. The smartest place to start is with energy efficiency, a friend at your side, and a specific date, time and objective on your calendar.

NOTES

1. It took a year and a half for me to start writing these pages.
2. "Remnant" indicates that the prairie soil and microbial life have never been disturbed by ploughs or grazing. Such lands have the potential for tremendous biological diversity when restored to full health. They also sequester carbon in the soil, and act as oases for endangered pollinator species like local bees.
3. Water efficiency is a form of energy efficiency, because it is electricity-intensive to pump water around a municipality. Additionally, gas and/or electricity is used to heat water for home use.
4. The International Energy Conservation Code is updated every three years; in states where the code is adopted, new buildings are built to even more stringent standards in 2015, 2018 and 2021. The biggest gains in efficiency, however, were made with the 2012 code at over 30% vs. 2009 (U.S. Department of Energy Office of Energy Efficiency and Renewable Energy, 2012).
5. Multi-family buildings.
6. Kansas City Metropolitan Energy Center is a nonprofit which provides resources, outreach and training to make alternative fuels and energy efficiency "commonplace."
7. Wherever two disparate construction materials come together (wood and concrete), there is the potential for a poor fit and thus conditioned air leakage.

8. For the residential sector, this includes electric, water, natural gas, propane or heating oil bills for heating, cooling, water heating, major appliances and electronics (telephone and external power adapters, computers, TVs and DVD players, home office equipment and small appliances) (Bailey, A., 2016, U.D. DOE).
9. The nucleus of Climate Action KC was formed by local elected officials concerned about climate change and committed to taking action, including legislation.
10. Mid-America Regional Council is a quasi-governmental group working on issues such as transportation that require coordination across a nine-county, two-state area comprising greater Kansas City.

LITERATURE CITED

American Council for an Energy Efficient Economy (ACEEE). (2016). *Report: "Energy Burden" on low-income, African American & Latino households up to three times as high as other homes, more energy efficiency needed.* Washington, DC: ACEEE.

Ariely, D. (2008). *Predictably irrational.* New York: Harper.

Bailey, A. (2016). *Breaking down the typical utility bill.* Washington, DC: United States Department of Energy. Retrieved from https://www.energystar.gov/products/ask-the-expert/breaking-down-the-typical-utility-bill.

Baldwin, J. (1961). *Nobody knows my name – More notes of a native son.* New York: Vintage International.

Bardhan, A., Jaffree, D., Kroll, C., & Wallace, N. (2013). Energy efficiency retrofits for U.S. housing: Removing the bottlenecks. *Regional Science and Urban Economics, 7,* 45–60. Retrieved from https://www.sciencedirect.com/science/article/pii/S016604 6213000677.

Baussan, D. (2015). *Social cohesion: The secret weapon in the fight for equitable climate resilience.* Washington, DC: Center for American Progress. Retrieved from https://www.americanprogress.org/issues/green/reports/2015/05/11/112873/social-cohesion-thesecret-weapon-in-the-fight-for-equitable-climate-resilience.

Burkhardt, J. (2019). Behavioral inertia: The low cost of doing nothing. Retrieved from https://www.linkedin.com/pulse/inertia-low-cost-doing-nothing-john-burkhardt-ph-d-.

Burns, D. D. (1989). *The feeling good handbook.* New York: William Morrow and Company.

Center for Ecotechnology. (2021). *Building better: Passive house.* Pittsfield, MA: Center for Ecotechnology. Retrieved from https://www.centerforecotechnology.org/building-better-passive-house.

City of Kansas City, Missouri. (2019). *Advance KC dashboard – Income and poverty.* Kansas City, MO: City Office of Economic Development. Retrieved from https://dashboards.mysidewalk.com/kcmo-advancekc/income.

Drehobl, A., & Ross, L. (2016). *Lifting the high burden in America's largest cities: How energy efficiency can improve low-income and underserved communities.* Washington, DC: American Council for an Energy-Efficient Economy. Retrieved from https://www.aceee.org/sites/default/files/publications/researchreports/u1602.pdf.

Fanara, A., Clark, R., Duff, R., & Polad, M. (2006). How small devices are having a big impact on U.S. utility bills. In *Presentation at the proceedings of the 4th international conference energy efficiency in domestic appliances and lighting,* June 21-23, 2006, London, United Kingdom. Retrieved from https://www.energystar.gov/ia/partners/prod_development/downloads/EEDAL-145.pdf.

Gawande, A. (2013). Slow ideas. *The New Yorker.* July 29, 2013 issue. New York: Conde-Nast, Inc.

Gifford, R. (2011). The dragons of inaction: Psychological barriers that limit climate change mitigation and adaptation. *American Psychologist, 66*(4), 290–302. Retrieved from https://psycnet.apa.org/record/2011-09485-005.

Granade, H. C., Creyts, J., Derkach, A., Farese, P., Nyquist, S., & Ostrowski, K. (2009). *Unlocking energy efficiency in the U.S. economy executive summary.* New York: McKinsey & Company. Retrieved from https://www.mckinsey.com/~/media/mckinsey/dotcom/client_service/epng/pdfs/unlocking%20energy%20efficiency/us_energy_efficiency_exc_summary.ashx.

Grossberg, F., Wolfson, M., Mazur-Stommen, S., Farley, K., & Nadel, S. (2015). *Gamified energy efficiency programs.* Washington, DC: American Council for an Energy-Efficient Economy. Retrieved from https://www.aceee.org/research-report/b1501.

Handel, S. (2016). What great apes teach us about emotion. *Morality, and Civilization.* Retrieved from https://www.theemotionmachine.com/what-great-apes-teach-us-about-emotions-morality-and-civilization/.

Harvard Joint Center for Housing Studies. (2017). *America's rental housing 2017.* Boston, MA: Harvard University. Retrieved from https://www.jchs.harvard.edu/sites/default/files/02_harvard_jchs_americas_rental_housing_2017.pdf.

Hofmeister, B. (2010). *Using social psychology to design market interventions to overcome the energy efficiency gap in residential energy markets* (pp. 10–18). Detroit, MI: Wayne State University Law School.

Holzwarth, A. (2019). *The three laws of human behavior.* London, UK: Behavioral Science Solutions, LTD. Retrieved from https://www.behavioraleconomics.com/the-three-laws-of-human-behavior/.

Hynek, D., Levy, M., Smith, B., & Wisconsin Division of Energy Services. (2012). *"Follow the money": Overcoming the split incentive for effective energy efficiency program design in multi-family buildings. Abstract.* Washington, DC: American Council for an Energy-Efficient Economy. Retrieved from https://www.aceee.org/files/proceedings/2012/data/papers/0193-000192.pdf.

Income by Zip Code. (2021). *64128 income statistics.* Austin, TX: Income By Zip Code, Inc. Retrieved from www.incomebyzipcode.com/Missouri/64128.

International Energy Agency. (2016). *Energy efficient prosperity: The "first fuel" of economic development.* Paris, France: International Energy Agency. Retrieved from https://www.iea.org/news/energy-efficient-prosperity-the-first-fuel-of-economic-development.

Jaffe, E. (2013). *Why wait? The science behind procrastination.* Washington, DC: Association for Psychological Science. Retrieved from https://www.psychologicalscience.org/observer/why-wait-the-science-behind-procrastination.

Kahneman, D., & Tversky, A. (1992). Advances in prospect theory: Cumulative representation of uncertainty. *Journal of Risk and Uncertainty, 5*(4), 297–323.

Kelly, M. (2013). *Poor concentration: Poverty reduces brainpower needed for navigating other areas of life.* Princeton, NJ: Princeton University. Retrieved from https://www.princeton.edu/news/2013/08/29/poor-concentration-poverty-reduces-brainpower-needed-navigating-other-areas-life.

Kouyoumdjian, H. (2012). Learning through visuals. *Psychology Today.* New York: Sussex Publishers, LLC.

Kuhns, M. R., & Miller, R. H. (2011). *Planting trees for energy efficiency: The right tree in the right place.* Provo, UT: Utah State University Forestry Extension, 2 pp. Retrieved from https://forestry.usu.edu/trees-cities-towns/tree-selection/plant-trees-energy-conservation.

Levy, J., Woo, M. K., Penn, S. L., Omary, M., Tambouret, Y., Kim, C. S., & Arunachalam, S. (2016). Carbon reductions and health co-benefits from US residential energy efficiency measures. *Environmental Research Letters, 11*(3). Philadelphia, PA: IOP Science Publishing. Retrieved from https://iopscience.iop.org/article/10.1088/1748-9326/11/3/034017.

Lovins, A. B. (2018). How big is the energy efficiency resource? *Environmental Research Letters, 13*(9), 2. Philadelphia, PA: IOP Science Publishing. Retrieved from https://iop-science.iop.org/article/10.1088/1748-9326/aad965.

Lyderson, K. (2014). Chicago group sells landlords on benefits of energy efficiency. *Energy News Network.* Retrieved from https://energynews.us/2014/11/04/chicago-group-sells-landlords-on-benefits-of-energy-efficiency/.

Mid-America Regional Council. (2015). *Regional multi-hazard mitigation plan, 4: Risk and vulnerability assessment.* Kansas City, MO: Mid-America Regional Council.

Molina, M. (2018). *Renewables are getting cheaper, but energy efficiency, on average, still costs utilities less.* Washington, DC: American Council for an Energy-Efficient Economy. Retrieved from https://www.aceee.org/blog/2018/12/renewables-are-getting-cheaper-energy.

National Churchill Museum. (2021). https://www.nationalchurchillmuseum.org/never-give-in-never-never-never.html

National Multi-Family Housing Council. (2021). *Household characteristics.* Washington, DC: National Multi-Family Housing Council. Retrieved from https://www.nmhc.org/research-insight/quick-facts-figures/quick-facts-data-download/.

National Rural Housing Coalition. (2021). *Housing need in rural America.* Retrieved from https://ruralhousingcoalition.org/overcoming-barriers-to-affordable-rural-housing.

North American Council of Insulation Manufacturing. (2015). *Ninety percent of U.S. homes are underinsulated.* Alexandria, VA: PRNewswire. Retrieved from https://www.prnewswire.com/news-releases/ninety-percent-of-us-homes-are-under-insulated-300151277.

Pandit, R., & LaBand, D. N. (2010). Energy savings from tree shade. *Ecological Economics, 69*(6), 1324–1329. Amsterdam, Netherlands: Elsevier, Inc. Retrieved from https://www.sciencedirect.com/science/article/abs/pii/S0921800910000340.

Ross, L., Drehobl, A., & Stickles, B. (2018). *The high cost of energy in rural America.* Washington, DC: The American Council for an Energy Efficient Economy. Retrieved from https://www.aceee.org/sites/default/files/publications/researchreports/u1806.pdf.

Schultz, W. P., Nolan, J. M., Cialdini, R. B., Goldstein, N. J., & Griskevicius, V. (2007). The constructive, destructive and reconstructive value of social norms. *Association for Psychological Science,* 18(5), 429–434.

Shift eLearning. (2014). *Studies confirm the power of visuals to engage your audience in learning.* San Jose, Costa Rica: Shift Disruptive eLearning. Retrieved from https://www.shiftelearning.com/blog/bid/350326/studies-confirm-the-power-of-visuals-in-elearning.

Sussman, R., Gifford, R., & Abrahamse, W. (2016). *Social mobilization: How to encourage action on climate change.* Victoria, BC: The Pacific Institute for Climate Solutions.

The Earth Institute at Columbia University. (2010). *Many Americans are still clueless on how to save energy.* New York: Columbia University. Retrieved from https://www.science-daily.com/releases/2010/08/100817103352.htm.

Turner-Skoff, J. B., & Cavender, N. (2019). *Benefits of trees for livable and sustainable communities.* New Phytologist Foundation. Hoboken, NJ: Wiley Online Library. Retrieved from https://nph.onlinelibrary.wiley.com/doi/full/10.1002/ppp3.39.

Underhill, L. J. (2018). *Energy efficiency, indoor air quality and health: A simulation study of multi-family housing in Boston, MA.* Boston, MA: Boston University. Retrieved from https://open.bu.edu/handle/2144/33047.

Underhill, L. J., Dols, W. S., Lee, S., Fabian, P., & Levy, J. I. (2020). Quantifying the impact of housing interventions on indoor air quality and energy consumption using coupled simulation models. *Journal of Exposure Science and Environmental Epidemiology, 30*(3), 1–12. Berlin, Germany: Nature Portfolio.

U.S. Census Bureau. (2019). *2019 Median household purchasing power for Missouri.* Suitland, MD: United States Department of Commerce.

U.S. Department of Agriculture Economic Research Service. (2021). *Rural poverty and well-being*. Washington, DC: U.S. Department of Agriculture Economic Research Service. Retrieved from https://www.ers.usda.gov/topics/rural-economy-population/rural-poverty-well-being/.

U.S. Department of Agriculture Forest Service, Pacific Southwest Research Station. (2011). *Trees pay us back—Urban trees make a good investment*. Washington, DC: U.S. Department of Agriculture Forest Service. Retrieved from https://www.fs.fed.us/psw/news/2011/110308_arborday.shtml.

U.S. Department of Energy, & Energy Information Administration. (2015). *EIA residential energy consumption survey, 2015*. Washington, DC: United States Department of Energy.

U.S. Department of Energy, & Energy Information Administration. (2017). *One in eight U.S. home uses a programmed thermostat with a central air conditioning unit*. Washington, DC: U.S Department of Energy. Retrieved from Eiagov/todayinenergy/detail.php?id=32112.

U.S. Department of Energy, & Energy Information Administration. (2019). *Use of energy explained: Energy use in homes*. Washington, DC: U.S. Department of Energy. Retrieved from https://www.eia.gov/energyexplained/use-of-energy/homes.php.

U.S. Department of Energy & U.S. Environmental Protection Agency. (2021). *Energy star portfolio manager*. Retrieved from https://www.energystar.gov/buildings/benchmark.

U.S. Department of Energy Better Buildings Program. (2021). *Energy treasure hunts*. Washington, DC: U.S. Department of Energy. Retrieved from https://betterbuildings solutioncenter.energy.gov/better-plants/energy-treasure-hunts.

U.S. Department of Energy Office of Energy Efficiency and Renewable Energy. (2012). *National energy and cost savings for new single- and multi-family homes: A comparison of the 2006, 2009, 2012 editions of the IECC*. Washington, DC: U.S. Department of Energy. Retrieved from https://digital.library.unt.edu/ark:/67531/metadc834033/.

U.S. Department of Energy Office of Energy Efficiency and Renewable Energy. (2021). *Lighting choices to save you money*. Washington, DC: U.S. Department of Energy. Retrieved from https://www.energy.gov/energysaver/save-electricity-and-fuel/lighting-choices-save-you-money.

U.S. Department of Energy Office of Energy Efficiency and Renewables. (2021). *Industrial assessment centers*. Washington, DC: U.S. Department of Energy. Retrieved from https://www.energy.gov/eere/amo/industrial-assessment-centers-iacs.

U.S. Energy Information Administration. (2015). *EIA residential energy consumption survey, 2015*. Washington, DC: United States Department of Energy.

U.S. Energy Information Administration. (2017). *One in eight U.S. home uses a programmed thermostat with a central air conditioning unit*. Washington, DC: U.S Department of Energy. Retrieved from Eiagov/tod ayine nergy /deta il.ph p?id=32112.

U.S. Energy Information Administration. (2019). *Use of energy explained: Energy use in homes*. Washington, DC: U.S. Department of Energy. Retrieved from https://www.eia.gov/energyexplained/use-of-energy/homes.php.

4 Climate Change in Grassland Ecosystems
Current Impacts and Potential Actions for a Sustainable Future

Jesse Nippert, Seton Bachle,
Rachel Keen, and Emily Wedel

CONTENTS

DOI: 10.1201/9781003048701-4

4.1 BACKGROUND

4.1.1 WHY SHOULD WE CARE ABOUT GRASSLANDS AND THE IMPACTS OF CLIMATE CHANGE ON GRASSLAND ECOSYSTEMS?

For millions of years, humans have relied on grassland ecosystems for our survival – grasslands are globally ubiquitous and support agricultural livelihoods and the global food economy, provide forage for domesticated grazers, are cultivated for biofuels and fiber and are key regulators of global hydrological and biogeochemical cycling. Despite the vital services grasslands provide, they are often overlooked in favor of other, more charismatic ecosystems. Grasslands are often presumed to lack the beauty and visual grandeur of mountainous regions, or the rich history of old-growth forests. Compared with tropical forests, grasslands might appear to have lower biodiversity or productivity. Grasslands are also typically presumed to lack the unique and fantastic physiological adaptations allowing organisms to survive in climatically extreme locations, as in the Arctic tundra or hot deserts in the subtropics. However, there are few ecosystems that provide as many critical services for human demands, or that have been as heavily impacted as grasslands.

As the climate changes, many of the key services provided by grassland ecosystems are threatened (Section 4.3). The importance of grasslands to human civilization might lead one to assume that assessing, forecasting and mitigating the consequences of climate change in grassland ecosystems would be of paramount interest. And yet, grasslands are typically among the last terrestrial ecosystems to rally conservation concern. Due to extensive global modification of grasslands, undisturbed or "natural" ecosystem states are uncommon. For this reason, it is impractical to assess how climate change impacts "undisturbed grasslands" because there are almost none of those areas remaining today. Thus, we must instead consider how climate change impacts grasslands in the context of disturbance – both natural disturbances, such as fire and periods of severe drought, and human-caused disturbances, such as habitat fragmentation and changes in land management.

Despite being highly modified for human use, grasslands are often resilient and responsive to management intervention. Grasslands can "bounce back" from small-scale or minor perturbations and return to pre-disturbance conditions (a response defined as high resilience). However, when modifications are severe in scope or continue for extended periods of time, grassland resilience is reduced and the likelihood of persistence in permanent, degraded conditions is high. The threats to grasslands posed by climate change are extensive, interactive and harbor the potential to negatively and irreparably (in the context of human lifespans) alter the species present and services provided by these ecosystems. For these reasons, an assessment of climate change impacts and potential solutions to mitigate grassland ecosystem degradation are imperative.

4.1.2 THE SCOPE AND STRUCTURE OF GRASSLAND ECOSYSTEMS

Grasslands are typically defined as "open ecosystems" characterized by high cover of grasses and other grass-like plant species, including sedges and rushes (Bond, W. J., 2019). This is in direct contrast with "closed ecosystems," in which tree canopies have sufficient density to restrict light from reaching the ground surface, preventing the development of an herbaceous understory layer. While many grassland ecotypes receive sufficient annual rainfall to support tree growth and the development of a closed ecosystem, frequent disturbance in the form of climate variability (drought, flooding and extreme temperatures), fire and grazing by large mammalian herbivores maintains a stable, open ecosystem. This unique attribute of grasslands results in the restriction of the classical ecological concept of succession toward a "climax community" (Bond, W. J. 2019). Given that grasslands are old ecosystems that predate hominin evolution (Strömberg, C. A. 2011), the view that grasslands are part of an ecological continuum toward a forest ecosystem has been thoroughly debunked (Veldman et al. 2015). Instead, grasslands exist as a stable ecosystem state, unique from woodlands or forests so long as key ecosystem drivers are present (Staver et al. 2011).

Data were derived from the ESRI World Terrestrial Ecosystems package *https:// landscape12.arcgis.com/arcgis/rest/services/World_Terrestrial_Ecosystems/ ImageServer*

Current estimates classify 31–40% of terrestrial surface as grassland (36.7 million km^2; Gibson, D.J. & Newman J. A. 2019) (Figure 4.1). Grassland ecosystems occur on every continent except Antarctica– it is the largest biome (in terms of pre-colonial acreage) in Africa and North America (~3 million km^2 each) and comprises large regions of South America and northern Australia (Dixon et al. 2014). The species composition and physiognomic structure of grassland ecosystems varies biogeographically, often according to global gradients in aridity and temperature as well as the intensity of top-down drivers, including fire and herbivory (Bond 2019). Grasslands are associated with the vegetative dominance by grass species, and can often include high grass biomass production. Characteristic grassland ecosystems include the Patagonian and Mongolian steppes, Eurasian meadows and grasslands of the Great Plains region of North America. Grasslands can also commonly contain mosaics of other plant types, including forbs (herbaceous, non-grass species) and succulents, as well as trees and shrubs in varying distributions. Grassland ecotypes with a substantial woody layer include the open-savanna Cerrado of South America, the dry steppes of Eurasia and the tropical and semitropical savannas of Africa and Australia. Thus, despite an initial apparent simplicity ("it's just grass"), the types of plant species and their distributions can vary markedly across grassland types.

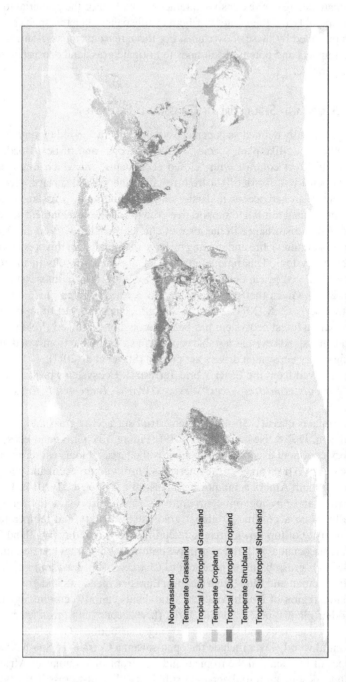

Nongrassland

Temperate Grassland

Tropical / Subtropical Grassland

Temperate Cropland

Tropical / Subtropical Cropland

Temperate Shrubland

Tropical / Subtropical Shrubland

FIGURE 4.1   Global distribution of temperate and tropical grasslands, shrublands and croplands.

These dynamics can also be highly variable within a grassland ecotype. High plant species richness and diversity is a common characteristic of grasslands around the world (Knapp et al. 1998; Blair et al. 2014). Dominant grass species coexist with a wide range of forbs, sedges, wetland plants and woody species. In addition to this large variety of plant functional types, grasslands also contain a variety of plant life-forms – depending on the grassland ecosystem, climate and evolutionary history, some grasslands are characterized predominantly by annual species (Mediterranean grasslands and grasslands in California), while others are predominantly peren-nial (African savannas and North American prairies). In short, all grasslands have a mosaic of species with varying life history and morphological and physiological attributes (Knapp et al. 1998; Blair et al. 2014). However, this high species richness reported is somewhat paradoxically accompanied by high species dominance by a few grass species. This pattern of high dominance accompanied by high overall diversity likely reflects the local interplay of multiple interacting drivers – fire, graz-ing and climate. These dynamics create increased niche space and locations where competition for resources is very high (where a few species will rise to dominance) as well as locations where only species with specialized adaptations to low resource availability or frequent disturbance can persist. The legacies of grazing systems and fire also impact plant species richness and dominance. For example, while frequent fire alone promotes the production of a few dominant grass species, grazing by large mammals reduces dominant grass cover and promotes forb establishment, increas-ing grassland plant diversity (Hartnett et al. 1996).

The high rates of growth and production by grasses in grassland ecosystems may be attributable to their unique morphology and physiology. As noted earlier in this chapter, many grasslands experience frequent disturbance as well as wide swings in short-term weather and longer term climate conditions. One key adaptation that benefits grass species is the location of their basal meristems, which is the site of new growth. Grasses position their meristems just below the soil surface, protecting these vital tissues from disturbance. This is one of the primary reasons that frequent fires actually promote grass growth rather than suppress it. After fire removes the litter and aboveground biomass, the bare soil surface quickly warms and the high-light environment facilitates rapid growth of new grass tissue from the protected meristems. In turn, grass production during the growing season provides biomass that acts as fuel for subsequent fire. In this way, frequent disturbance effectively maintains these open, diverse, grass-dominated ecosystems (Blair et al. 2014).

4.1.3 Ecosystem Services Provided by Grasslands

Grasslands provide many key ecosystem services, defined as the direct or indirect benefits of healthy environments to humans and societies. The most obvious direct ecosystem service provided by grassland ecosystems is their key role in global food security. Conversion of grassland to cropland underlies modern agriculture, and the direct, positive benefits to humans by this conversion is obvious. However, roughly 70% of global grassland area and 50% of global savanna area have been converted to agricultural land (Ramankutty et al. 2008), and these conversions often result in a loss of most indirect grassland ecosystem services (see Section 4.3 for a more detailed discussion of this topic). In addition to conversions to cropland, sustainable management of rangelands and grasslands provides pasture for domesticated livestock and supports a grazing economy of US$63 billion/year in the United States alone (Allred et al. 2014). This ecosystem service benefits ranchers and pastoralists from diverse walks of life in both first and third world countries and takes place in nearly all grassland ecosystems worldwide.

Some of the primary indirect ecosystem services provided by grasslands include (1) sequestration of carbon (C) belowground; (2) regulation of the water cycle; and (3) provisioning of habitat for invertebrates (including pollinators), wildlife and humans. Grassland plant species typically allocate more biomass belowground than aboveground (Gibson 2009), and these dense root systems contribute to the development of soil organic C pools over time, effectively removing C from the atmosphere. Grassroot systems are also adept at maintaining soil structure and minimizing erosion during high-intensity rain events. Grassland plant species regulate the water cycling by minimizing runoff, facilitating water infiltration within the soil profile and ultimately recycling this moisture back to the atmosphere via transpiration. The varying physiognomy of grasslands, along with high plant species diversity, provides mosaics of habitat and benefits for a wide range of animal species. A key trait of many grassland ecosystems is the functional redundancy provided by similar species (e.g., different grass species providing similar ecological roles), which allows for a robust maintenance of these ecosystem services as populations of specific species increase and decrease in response to disturbance through time.

Climate change impacts these direct and indirect ecosystem services by threatening the unique characteristics of grassland ecosystems. One might presume that effective adaptation to disturbance – a key grassland trait – posits that grasslands should be highly buffered against alterations caused by climate change. Indeed, grasslands are well adapted to disturbance, but the changes associated with a warming climate are often interactive and are occurring faster than historical changes in climate (USGCRP 2018). When these changes result in alterations to grassland vegetative structure, they can result in degraded lands, shifts to alternative ecosystem states (woodland or forest) or loss of habitat to invasive species. In Section 4.2, we will explain the direct consequences of climate change on grassland ecosystems, including how increasing atmospheric CO_2 concentrations, increasing global temperatures and altered rainfall patterns impact the growth dynamics and competitive relationships among grassland species. In Section 4.3, we will discuss several indirect consequences of climate change, primarily due to interactions with changes in land-use or land-cover mediated by humans. Finally, in Section 4.4, we highlight many of the potential climate solutions that we can participate in now, which could provide both smaller and greater remedies to offset the negative consequences of climate change on one of Earth's greatest ecosystems.

4.2   DIRECT CLIMATE CHANGE IMPACTS

Natural changes in the Earth's climate have occurred on the scale of hundreds to millions of years throughout geologic time. This natural variability is influenced by large-scale events, including predictable changes in the Earth's tilt and orbit around the sun (Milankovitch cycles), fluctuations in the intensity of solar radiation reaching the Earth, movement of tectonic plates and volcanic eruptions. Human activity, particularly the burning of fossil fuels since the onset of the Industrial Revolution, has resulted in sharp deviations from those natural, long-term climate dynamics. The buildup of greenhouse gases – CO_2, methane and nitrous oxide, among others – in the atmosphere creates a "greenhouse effect" by absorbing long-wave radiation (heat) emitted from the Earth's surface that would otherwise pass through the atmosphere. This trapped energy heats the atmosphere and has resulted in increasing global mean temperatures through time. Current atmospheric CO_2 concentrations are nearly 420 parts per million (ppm), whereas concentrations in the mid-1700s were roughly 280 ppm – this is nearly a 70% increase in only a few hundred years. As a result, mean global surface temperatures have increased by ~1.14°C since 1880 and are projected to increase another 1–6° by the end of the century. Because warm air is able to hold more moisture than cooler air, this increase in air temperature also impacts global rainfall patterns and is expected to facilitate more variable and extreme precipitation regimes. Changes at this scale have both direct and indirect impacts on ecosystems around the world, affecting water availability, growing season lengths, plant productivity and phenology and global nutrient cycling. However, climate change will not impact all plant species or ecosystems in the same way. For example, some species may respond positively to increased CO_2 concentrations, but simultaneously respond negatively to increased temperature and rainfall variability. Forecasting the consequences of climate change on grassland ecosystems requires an understanding of how key plant species respond to changes in CO_2 concentration, temperature and rainfall variability, and how shifts in their abundance impacts soil communities, C-cycling and nutrient fluxes.

4.2.1 ATMOSPHERIC CO_2 CONCENTRATIONS

CO_2 is the inorganic C substrate required for photosynthesis and the starting point for the vast majority of the complex organic molecules synthesized on Earth. CO_2 is generally well-mixed in the atmosphere and present in similar concentrations around the world on an interannual timescale. There are local and temporal variations that reflect differences in seasonality between the two hemispheres, the impact of increased emissions associated with urban environments and the ecological differences reflecting varying C-assimilation and respiration rates among different ecosystem types. Increased CO_2 concentrations typically correspond with increased rates of photosynthesis so long as other resources (e.g., soil moisture, light and nutrients) are available and environmental conditions promote growth. Therefore, it follows that rising atmospheric CO_2 concentrations would lead to increased rates of plant growth and global primary productivity. This response is often referred to as the "fertilization effect" and is typically viewed as a potentially positive outcome of rising greenhouse gas emissions. As noted previously, this fertilization effect hinges on the availability of other plant-limiting resources – especially water (Körner 2006).

To investigate the impacts of rising CO_2 on plant growth, experiments began in the 1980s, both in laboratory and in greenhouse settings, or in natural environments and outdoor experiments, for a large number of terrestrial ecosystems – forests, grasslands, tundra and semiarid desert, in both temperate and tropical regions. Many syntheses and reviews that detail the outcomes of these experiments have been previously published. While most of the field-based, or "natural," CO_2 fertilization experiments did exhibit increased rates of photosynthesis and growth in the short term, the longer-term consequences often varied from original predictions and illustrated how ecological interactions mediate the fertilization effect and vary among plant species, ecosystem type and geographic region (Körner 2006).

One major characteristic that modifies a plant species response to changes in atmospheric CO_2 concentrations is the specific photosynthetic pathway utilized. Over 85% of the plant species on Earth use the C_3 photosynthetic pathway. For these plant species, CO_2 assimilation reflects the concentration of CO_2 in the atmosphere (supply) and the concentration of CO_2 inside of the leaf (demand). Leaf internal CO_2 concentrations are largely regulated by the small pores that exist within the leaf surface that open and close to allow diffusion of air into the leaf and, simultaneously, water out of the leaf. Thus, as the supply of CO_2 in the atmosphere increases, the rate of C-assimilation in C_3 plant species increases (Pearcy and Ehleringer 1984). Within our tropical and temperate grasslands, many grass species use an alternative photosynthetic pathway, referred to as C_4 photosynthesis. C_4 plant species have a unique morphological adaptation within leaves that keeps CO_2 concentrations very low in the leaf internal spaces, which increases CO_2 demand, while simultaneously concentrating CO_2 in specialized bundle sheath cells containing the enzyme responsible for C-fixation. The outcome of this spatial separation of photosynthesis is the high rates of photosynthesis common to C_4 grass species (Pearcy and Ehleringer 1984). Given that photosynthesis in C_4 grasses is already occurring in a high-CO_2 environment (bundle sheath cells) near maximum enzymatic capacity, these species were not expected to show a growth response to increased atmospheric CO_2 concentrations compared to coexisting C_3 plant species. While this prediction typically holds when growing conditions are optimal (as in greenhouses or growth chambers), experiments in natural grasslands often show conflicting or non-intuitive results (Reich et al. 2018). These results illustrate how multiple resource limitations– including non-optimal soil moisture, light or nutrient availability – modify resource interactions to complicate ecological predictions based on increasing atmospheric CO_2 concentrations.

How did many C_4 grass species defy initial physiological predictions of being insensitive to increased atmospheric CO_2, and in some cases even perform better than coexisting C_3 plant species? The answer requires an understanding of how ecological interactions vary according to multiple resource limitations. As previously mentioned, grasslands are characterized by periods of low water availability resulting in dormancy or periods of low growth rates. When water availability is low, plants must reduce their photosynthetic rates in order to reduce water loss via transpiration – if leaf pores remain open to allow for photosynthesis to continue during drought conditions, the plant risks desiccation. Although this physiological process is driven by soil water conditions, it reduces rates of photosynthesis because CO_2 uptake and water loss occur through the same pores on the leaf surface. Even if atmospheric concentrations are high, photosynthesis remains low if the leaf pores are partially or fully closed due to low soil water conditions. C_4 grass species benefit more than C_3 species under conditions of low water availability and high CO_2 concentrations based on the previously described supply–demand dynamics of CO_2. C_4 grasses have a lower internal leaf CO_2 concentration compared to C_3 grasses due to their ability to concentrate CO_2 in the bundle sheath cells. Therefore, the driving gradient for CO_2 diffusion from the atmosphere into the leaf internal spaces is larger in C_4 species, even under low soil water conditions when resistance to diffusion through leaf pores is high (Dijkstra et al., 2010). Thus, these experiments have shown that the theoretical impacts of elevated CO_2 can deviate from real-world conditions, and also illustrated that global warming conditions that result in increased intensity or frequency of drought have the capacity to counteract any positive consequences of increased CO_2 on plant photosynthesis and grassland primary productivity.

4.2.2 TEMPERATURE

Surface air temperatures have increased in grasslands worldwide as a consequence of global warming, and this trend is expected to continue throughout the century, with increases dependent upon varying global warming projections. There are two unique features of many grasslands and savannas worldwide that could, in theory, mute the negative consequences of future warmer air temperatures on grassland species and ecosystem processes. First, many grasslands currently occur in regions that have historically experienced periods of high air temperature. These periods may occur predictably during the summer growing season or can be associated with interannual climate anomalies such as El Niño or La Niña conditions. This history of periodic drought has resulted in grasslands and savannas being well adapted to periods of low water availability. Second, for many grassland ecosystems, the dominant grass species utilize the C_4 photosynthetic pathway. The optimal temperature conditions for C-assimilation in C_4 species are higher (33–37°C) than species utilizing the C_3 photosynthetic pathway (28–32°C) (Sage and Kubien 2007). For this reason, higher air temperatures could be expected to have fewer negative impacts on physiological functioning in C_4 species. However, as has been reported multiple times in this chapter, the drivers of climate change do not occur in isolation, and the dynamics of ecological communities have complex responses to interactive environmental factors. For these reasons, increased air temperatures as a consequence of global warming are expected to have several negative consequences for grassland ecosystems.

The most proximal direct effect of warmer air temperatures is increased leaf energy budgets and increased cellular and soil respiration rates. Increased air temperatures alter the energy budgets of leaves by changing energy dissipation pathways (both sensible and latent heat exchange) and energy transfer. Reduced energy exchange with the atmosphere increases leaf temperature and leads to greater potential of physiological stress. Cellular-level stress increases because C-assimilation is enzymatically driven, and enzymatic processes operate within certain temperature conditions (Sage and Kubien 2007). Exceeding optimum temperatures for extended periods of time increases leaf stress and typically results in reduced photosynthetic rates and growth. Increased air temperature also impacts C-assimilation through tighter regulation of leaf gas exchange. As temperatures increase, the vapor pressure deficit that exists between a leaf and the atmosphere increases, resulting in higher rates of leaf water loss for a given stomatal (leaf pore) aperture. Thus, grassland species have to reduce rates of gas exchange, and therefore reduce CO_2 uptake, in order to minimize the risk of desiccation. Finally, increased air temperature directly increases cellular and soil respiration rates. For every 10°C increase in temperature, cellular respiration rates double – a phenomenon referred to as a Q10 response (Tjoelker et al. 2001). Consequently, increased air temperatures in grasslands speed up the rate of C-cycling for both vegetation and soil communities, ultimately leading to a reduction in the amount of C stored by the ecosystem (Figure 4.2).

In addition to the potential for increased physiological stress associated with higher air temperatures, there are many indirect community-level consequences. Increased annual air temperatures are resulting in changes in the length of vegetation growing seasons, especially for temperate grassland regions where the dormant season is associated with cold temperatures. As temperatures warm, the growing season shifts earlier in the year and typically lasts longer. These shifts in season can result in phenological mismatches between plants and their pollinators or herbivores, or between periods when fires occur naturally and when plant species are physiologically less sensitive to fire (dormant vs. actively growing vegetation). Additionally, an earlier initiation of spring growth, coupled with a longer overall growing season and

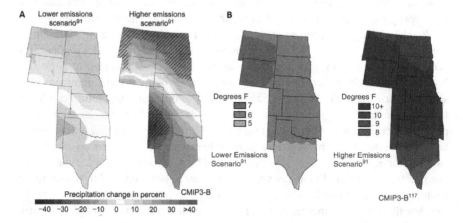

FIGURE 4.2 Forecast changes in (A) precipitation and (B) air temperature for the Great Plains region of the United States by the end of the century. Compared to baseline data in the region from 1960 to 1979, precipitation change is projected to increase in the northern Great Plains, corresponding with a decrease in the southern Great Plains for both emission scenarios. Air temperature is expected to increase across the entire region, with the largest increases in northern states. *Image credit: U.S. Global Climate Change Impacts in the United States: A State of Knowledge Report from the U.S. Global Change Research Program*, editors: Karl, T. R., J. M. Melillo, and T. C. Peterson. 2009.

later fall senescence, will likely result in larger reductions in available soil moisture and more pronounced growing season drought effects. Even with no reduction in annual precipitation amount, longer growing seasons result in larger annual evapotranspirative fluxes, greater depletion of stored soil moisture and increased physiological stress on plants. If these stresses accumulate early in the growing season, late-season flowering plant species may experience more frequent disruptions in their life cycles. So far, this discussion of elevated air temperature modifying local ecohydrology has assumed no overall changes in magnitude of annual precipitation amount, timing or intensity, but the next section will detail the expected changes to grassland precipitation patterns as the climate warms.

4.2.3 PRECIPITATION VARIABILITY

In addition to increasing global mean air temperatures, climate change is expected to have substantial impacts on the global hydrologic cycle (Giorgi et al. 2019). Most model projections agree that precipitation variability will increase as the climate continues to warm (Pendergrass et al. 2017), likely resulting in longer periods of drought punctuated by more extreme precipitation events. Globally, water is the main limiting resource for plants, and precipitation and soil water availability are some of the major determinants of biome distributions. Both the magnitude of total annual precipitation and the timing and size of precipitation events are often critical in maintaining ecosystem dynamics and function, and grasslands are no exception to this trend.

Changes in precipitation patterns associated with climate warming will not be uniform in all grassland ecosystems. Grasslands located in midlatitude or subtropical dry regions are expected to experience a net decline in annual precipitation, while grasslands at higher latitudes have mixed predictions regarding changes in total annual amount (Gibson and Newman 2019). While the magnitude of change in precipitation will vary regionally, general circulation models (GCMs) consistently predict that intra-annual variability in precipitation will increase as the climate warms (IPCC 2007). Productivity in grassland ecosystems is largely impacted by inter- and intra-annual precipitation patterns (Knapp et al. 2016). Low annual precipitation is associated with lower aboveground net primary productivity (ANPP) (Nippert et al. 2006), owing in part to lower soil moisture availability and increased plant water stress resulting in lower photosynthetic rates. However, increasing duration of drought events, even when total annual precipitation does not change, can decrease grassland productivity to a similar degree as low annual rainfall (Knapp et al. 2002). Less frequent, more intense precipitation events can also reduce infiltration of water into the soil and increase runoff, erosion, flooding and leaching of nutrients from the soil.

The impact of increased rainfall variability on grassland productivity is also expected to vary based on local climate conditions (Heisler-White et al. 2009). Experimentally altering precipitation frequency while maintaining overall amount of growing season precipitation (i.e., fewer, but larger, rainfall events) has been shown to increase productivity in semiarid grasslands (Heisler-White et al. 2008), while decreasing productivity in more mesic grassland ecosystems (Fay et al. 2003).

4.2.4 Special Grassland Example: Impacts of Increasing Precipitation Variability in Tallgrass Prairie

The North American tallgrass prairie exists within the mesic temperate biome of the central Great Plains (Hayden 1998). This grassland region is characterized by high grass productivity, driven in part by a climate regime that includes both warm growing-season temperatures and rainfall inputs that typically exceed losses from evapotranspiration (Briggs & Knapp 1995; Nippert et al. 2006). Historically, ~75% of rainfall events occur between March and September each year. Periods of low rainfall resulting in drought are characteristic of this region. These droughts can result from years with below-average total precipitation, or extended periods without rainfall within the growing season without a change in the total annual amount (Knapp et al 2002).

Many long-term experimental precipitation manipulations have been conducted in tallgrass prairie, starting in the mid-1990s. These experiments were designed to test grassland responses to multiple predictions of precipitation change forecast for tallgrass prairie, and mesic grasslands more broadly. To date, these experiments have included three main rainfall manipulations: (1) passive, but chronic, reduction in total annual rainfall amount with no change across years; (2) change in precipitation variability whereby no change in total annual rainfall amount, but rain events are larger and less frequent, resulting in longer intervals of dry days between larger rain events; (3) reductions in total rainfall amount for multiple consecutive years, followed by multiple consecutive years without rainfall reductions to investigating drought legacies (Figure 4.3); and (4) factorial designs. On occasion, these rainfall manipulations have included passive infrared air temperature increase of a few degrees or the simulation of heat waves, which can accompany drought in this region.

Several key results can be inferred from these precipitation experiments. As expected, reductions in total amount of annual rainfall reduces surface soil moisture availability and annual grassland biomass. The entire herbaceous community experiences reduced plant growth, with aboveground plant biomass typically having larger biomass reductions compared to belowground (Smith, 2011; Wilcox et al. 2017).

FIGURE 4.3 The climate extremes experiment at the Konza Prairie Biological Station. Experimental manipulations of precipitation using large rain-out shelters allow ecologists to test the predictions of climate change on intact grassland communities. *Image credit*: Melinda D. Smith.

Changes in precipitation variability, without changes in the total amount of rain delivered per year, have resulted in chronic droughts, reduced biomass, altered soil respiration fluxes and changes in species composition (Knapp et al. 2002, Fay et al. 2011). Precipitation manipulations that sequentially expose grasslands to years of reduced rainfall followed by recovery illustrate a high resilience (e.g., reduced biomass and canopy cover during drought, but recovery to pre-drought conditions in subsequent years) of the C_4 grass species to these changes, but reduced resilience by the coexisting forb communities (Hoover et al. 2014). In total, these experiments illustrate that this mesic grassland is highly responsive to changes in both precipitation amount and timing. Future changes in rainfall are highly likely to impact both ecosystem function with regard to C uptake and cycling, as well as the abundance and cover of the plant species in this grassland community (Felton et al. 2019; Knapp et al. 2020).

As discussed earlier in this chapter, changes in atmospheric CO_2 concentration, increased air temperatures and changes in precipitation pattern and amount are likely to have singular and interactive effects on grassland processes. The ability of grasslands to maintain their structural (number and proportion of species present) and functional (ecosystem services) processes depends on the maintenance of ecosystem resilience. Resilience describes the ability of an ecosystem to maintain stability and critical services despite perturbations (i.e., direct or indirect climate change impacts). In Figure 4.4, we summarized the likely impacts of the direct climate change discussed on grassland and rangeland. While local impacts will vary from place to place, cumulatively we can expect these climate change drivers to negatively impact grassland structure and function. Perhaps most alarming, the direct climate change impacts on grasslands are likely to exacerbate the indirect threats to grasslands – namely, woody encroachment and the spread of invasive plant and insect species. In Section 4.3, we detail these indirect impacts, explain why these impacts threaten grassland ecosystems and describe the potential acceleration of conversion from grassland to degraded land under future climate scenarios.

4.3 SECONDARY (INDIRECT) IMPACTS

4.3.1 LAND-USE/LAND-COVER CHANGE

In addition to the direct impacts of climate change on grassland ecosystem dynamics, these changes also indirectly impact grasslands via interactions with human-driven modifications in land-use and land-cover. Here, land-use and land-cover changes refer to anthropogenic modifications of the landscape resulting in an alteration of goods and services compared to a natural grassland ecosystem. In many parts of the world, the first lands cultivated for agriculture were those with relatively flat topography in productive grassland ecosystems. With increased population growth and subsequent demand for food, less productive grasslands and remnant fragments have been converted to agriculture, resulting in widespread reductions in the spatial extent of native grasslands. Humans also continue to drive reductions in the quality of remaining grasslands through modified fire regimes (too frequent or infrequent),

	↑ CO₂	↑ Temperature	↑ Precip. Variability	Interactions/ Net Effects
Native Grassland Stability	— Cancels out	→ More drought/ metabolic stress	→ More drought stress, changes in infiltration dynamics	→ Interactions between climate drivers will decrease grassland resilience
Agriculture/Rangeland Productivity	← Fertilization effect	→ Photorespiration, More drought stress	→ More drought stress, Incr. need for irrigation	→ Potentially faster growth, but likely offset by increased stress and declining crop/forage productivity
Woody Encroachment	← Fertilization effect	→ Photorespiration	← Increased infiltration	← Woody species benefit more than grass species with higher CO2 and greater water infiltration to depth
Spread of Invasive Species	← Fertilization effect	→ Favors C₄ invasives	← Decrease native species growth	← Direct effects that reduce stability of native grasslands will make it easier for invasive species to spread

FIGURE 4.4 Predicted impacts of direct climate change on grassland processes (stability, productivity – discussed in Section 4.2) and indirect climate change impacts (woody encroachment and invasive species – discussed in Section 4.3).

overgrazing, introduction of exotic and/or invasive species and fragmentation of habitat.

4.3.1.1   Agriculture, Urbanization and Habitat Fragmentation

Approximately one-third of Earth's terrestrial surface is used for agriculture (4.8 billion ha; Food and Agriculture Organization of the United Nations (FAO) report), and native grasslands historically comprised a large majority of this area. The original conversion of grasslands rather than forests to annual agriculture occurred primarily because (1) it is generally easier to replace a diverse grassy community with crop species, many of which are also grasses, and (2) land clearing of forests is laborious, intensive and still results in soils that retain large woody roots and stumps that make cultivation difficult. Indeed, conversion to agricultural land is the primary threat of grassland ecosystems globally (Gibson 2009). Initially, the first grasslands converted to agriculture were locations with the most fertile soils. With increased human population growth, demand for agricultural commodities increased, requiring increased production. Without more highly fertile lands to be converted to agriculture, the only options available were increased yields on existing lands through genetic breeding of key cultivars, and by the conversion of marginal grasslands (e.g., locations with less productive soils, or lower yield potentials) to agricultural land. Both of these scenarios are outcomes of increased industrialization of agriculture to support human demand, and both scenarios have resulted in increased losses of grasslands globally. Importantly, when marginally productive (less fertile) grasslands are converted to agriculture, more land is required to produce agricultural yields that feed the same number of people.

A secondary impact of grassland conversion to agriculture is landscape fragmentation. As the landscape is cultivated, small portions of the original landscape remain in isolated, small-acreage patches. This fragmentation reduces habitat availability for animal populations, results in the loss of natural grassland corridors that facilitate movement on the landscape and is particularly impactful on highly mobile populations of birds or migratory animals. Fragmentation also reduces genetic diversity of plant populations by reducing plant species richness and abundance within fragments and lowering the total number of species present within a given area (Krauss et al. 2010). With continued human population growth and increased migration of humans from rural to urban environments, the urban/suburban expansion is typically at the expense of agricultural and grassland regions that surround cities. These former grasslands are developed into urban environments or become degraded lands supporting low-income human populations in regions with large wealth disparities.

Expected future changes in climate, detailed in Section 4.2, increase the uncertainty of agricultural production and are a distinct threat to food security. Many key commodity species grown in grassland regions, like wheat, soybeans and barley, utilize the C_3 photosynthetic pathway and exhibit decreased production under warmer air temperatures (Ainsworth and Ort 2010), irrespective of irrigation and soil water status. The combination of warmer temperatures and increased precipitation variability is expected to further increase the likelihood of crop failure and lower yields for many agricultural species. The impacts of climate change on human livelihoods are not globally uniform, however. Climate change will disproportionately affect people living in grassland regions that already face malnourishment due to chronic poverty and income-based obstacles to sustainable agricultural intensification (e.g., sub-Saharan Africa and South-East Asia). Compounding the severity of this issue, these areas are also predicted to have the highest rates of population growth over the remainder of this century, resulting in rapidly increasing demand for a reduced supply of food. Beyond crop production, domesticated livestock depend on grasslands for forage. Future changes in climate are expected to reduce grass productivity, resulting in less forage available for livestock. Reduced available forage increases the potential for animal stress and the likelihood of livestock mortality, particularly if temperature increases and drought become more frequent (IPCC 2019).

For these reasons, climate change is likely to threaten global food security. However, modern agriculture could potentially be further facilitating changes in climate (Tilman et al. 2011), particularly through increased methane emissions and disruption of natural nitrogen cycling. In addition, high-yielding lands for sustainable agriculture are no longer available, and yet the future demand for larger amounts of food in the context of an uncertain climate is increasing. Future efforts must focus on novel methods of sustainable agriculture that can produce higher yields from existing agricultural land, reduce chemical inputs leading to reductions in soil quality or environmental contamination and reduce food waste and the disparity in food security that exists across nations and regions.

4.3.1.2 Woody Encroachment

The conversion of grasslands to agricultural land has led to a rapid and substantial decrease in grassland area worldwide. The fragmentation of remaining grasslands – due to continued conversion to agriculture, urbanization and changes in land-use – has facilitated an increase in woody plants in historically grass-dominated areas. This phenomenon is referred to as woody encroachment, and it is occurring in grasslands and savannas around the world. In many cases, the proliferating woody species are native to the region, but were historically restricted to riparian zones or have a patchy distribution throughout the landscape, as in savannas. Increased woody cover has led to a direct loss of grasslands and their ecosystem services – namely, quality forage for livestock grazing – by reducing herbaceous productivity and species richness (Archer et al., 2017). In addition, woody encroachment directly alters C and water cycling by shifting C-storage from primarily belowground (in soils and grass roots) to aboveground (in woody tissues) and increasing rates of evapotranspiration (O'Keefe et al. 2020) (Figure 4.5).

The primary drivers of woody encroachment are complex, and vary by grassland type (e.g., semiarid vs. mesic, or temperate vs. tropical). The amount of rainfall received each year establishes the maximum potential for tree or shrub cover in a grassland or savanna ecosystem, whereby potential woody cover increases with mean annual precipitation (Sankaran et al. 2008). The amount of annual rainfall largely determines the drivers that suppress woody cover. For example, in arid and semiarid grasslands, woody cover is primarily limited by precipitation – when precipitation is too low to support woody vegetation, woody cover will remain low. In more mesic grasslands that receive enough precipitation to support a higher abundance of woody

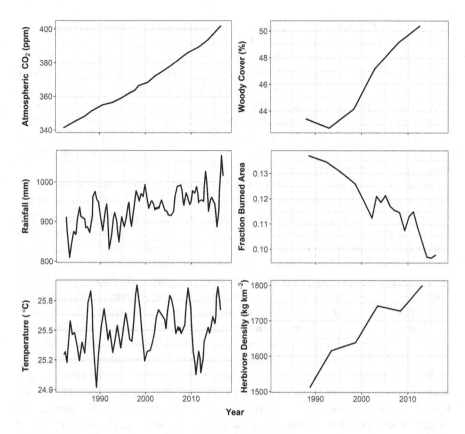

FIGURE 4.5 Changes in atmospheric CO_2, rainfall, air temperature, woody cover, fraction burned cover and herbivore density for sub-Saharan Africa over 1985–2015. Woody cover has increased over the past three decades, corresponding with increases in many environmental variables (CO_2, rainfall, air temperature and herbivore density), along with an overall decrease in the fraction burned area. *Source*: Figure modified from Venter et al. 2018.

plants, woody cover is limited by frequent disturbance – namely, fire and herbivory (Archer et al. 2017). Browsing and frequent fire suppress the spread of woody species by killing or injuring tree saplings and small shrubs, preventing them from establishing and maturing. When fire is frequent, it is sufficient to keep woody species in this vulnerable zone, but when fire frequency declines, woody plants have more time between fire events to establish and grow. If young trees are able to grow tall enough to escape the impacts of fire, or if shrubs grow large enough to shade out grasses, reduce fine fuels and fire intensity and prevent fire from damaging stems in the shrub interior, these woody plants can escape the "fire trap" (Ratajczak et al. 2014; Archer et al. 2017). Often, decreased fire frequency, decreased browsing and/ or overgrazing at the local level are considered the predominant drivers of woody encroachment. However, interactions with changing climate conditions are likely to exacerbate this process.

Increased atmospheric CO_2 concentration is expected to speed up woody plant establishment and growth – more CO_2 in the atmosphere often leads to increased carbon fixation via photosynthesis and can subsequently reduce water loss because stomatal conductance typically decreases with increasing [CO_2]. This "fertilization effect" is likely to be most beneficial to juvenile woody plants, which are the most vulnerable to disturbances such as fire and browsing. Increased growth rates facilitated by increasing CO_2 concentrations would accelerate the transition from juvenile to adult size classes in woody species, increasing their survival after disturbance. Additionally, greater carbon gain increases carbon storage in woody tissues that is used to produce new tissues following fire and browsing.

In addition to elevated atmospheric CO_2 concentrations, shifts in regional precipitation patterns are expected to differentially affect woody and herbaceous plants and further facilitate woody encroachment in grassland ecosystems. As discussed previously in this chapter, many grasslands are expected to experience more extreme and prolonged droughts punctuated by periods of unusually heavy rainfall as precipitation variability increases. These predicted changes in precipitation patterns are expected to benefit deep-rooted woody plants over shallow-rooted grasses (Kulmatiski and Beard, 2013). Deep roots give access to deep water sources that provide a consistent water source during dry periods and reduces woody plant competition with grasses for shallow soil water. Predicted increases in the intensity of rain events and magnitude of winter precipitation may increase soil water infiltration, recharging water in the deep soil layers and disproportionality benefitting deep-rooted woody plants over grasses.

Woody encroachment is a major risk to grassland ecosystems because the consequences are long-lasting and difficult to manage. Infrequent or lack of burning leads to a gradual grassland-to-woodland transition (Briggs et al. 2005; Bond 2019). Several lines of evidence suggest that the spread of woody vegetation in grassland ecosystems can reach a point of hysteresis, where it is impossible (or exceedingly difficult) to reverse this grassland-to-woodland transition (Bestelmeyer et al. 2011). Once woody vegetation has become established in the absence of fire, reimplementation of frequent fire is typically not sufficient to reverse the transition and restore grass cover (Staver et al. 2011; Miller et al. 2017). Spread of woody vegetation typically decreases fire frequency as surface fuel loads (grasses) are reduced and grass cover becomes patchier (Drewa and Havstad, 2001). This decrease in fire frequency further promotes the spread of woody vegetation, resulting in a positive feedback loop. Although woody encroachment in grasslands leads to lower frequency of surface fires, it can also result in higher fire intensity when wildfires do occur. This represents a shift in fire regime from the frequent surface fires that maintain open grasslands to less frequent, but more intense crown fires when woody vegetation does ignite, typically during periods of drought (Brooks et al. 2004).

4.3.1.3 Specific Grassland Example – Woody Encroachment in African Lowveld Savannas

The distribution of savannas is determined by a seasonal climate, occurring in locations with hot, wet summers and cooler, dry winters. The cycle of the wet and dry seasons has shaped the savanna ecosystem by driving the distribution and migration of large mammals and promoting hot fires after the dry season. These climate and land-use drivers have complex interactions maintaining an open canopy with low tree: grass ratios. African savannas are unique in that they span a large precipitation gradient (150–1,200 mm mean annual precipitation) and host a large diversity of mammals that feed on grasses (grazers) and woody plants (browsers). While precipitation determines the amount of woody plants the system can support, herbivory and fire suppress woody plant growth and kill woody saplings. It is hypothesized that semiarid savannas are stable systems where low water availability (<650 mm yr[1]) maintains low tree:grass ratios (Sankaran et al. 2008). In contrast, mesic savannas are reliant on frequent fires to maintain low tree abundance. Mesic savannas are considered particularly vulnerable to woody encroachment because of increased water availability and are reliant on frequent and intense fires to maintain low tree abundance. However, woody encroachment is occurring across the precipitation gradient in lowveld savannas, including mesic savannas that experience historical fire frequencies, suggesting global drivers likely interact with local changes to fire frequency and herbivore abundance (Stevens et al. 2016; Case and Staver 2017).

Africa hosts the world's last remaining intact megaherbivore guilds and many of the remaining large predators. The degradation of savannas and loss of biodiversity due to woody encroachment directly conflict with conservation goals to protect and conserve remaining ecologically intact savannas. Conservation areas, including National Parks and private reserves, rely on ecotourism to fund these protected areas. Increased tree and shrub cover is likely to negatively affect ecotourism as animals become harder to see with high woody cover (Gray and Bond 2013). Additionally, loss of habitat and forage may have cascading effects on the grazers and other mammals that rely on open, grassy systems. These effects may include a restructuring and redistribution of mammal communities as obligate grazers are lost and browsers become more frequent in encroached areas (Smit and Prins 2015). Management at the local level is likely required to combat encroachment, including frequent prescribed fire and bush clearing. Although these techniques can be expensive and time- and labor-intensive, they are likely required to combat woody encroachment in the face of global drivers (Figure 4.6).

Impacts of woody encroachment on grassland hydrology vary, depending on the local climate and geomorphology, the types of species present and the local land-use history. Due in part to their access to deeper, more consistent water sources, woody species typically have much higher rates of transpiration than grasses (O'Keefe et al. 2020), leading to greater overall water flux and potentially depletion of deeper soil water over time (Acharya et al. 2017). Transpiration in woody encroached areas has the potential to exceed precipitation inputs during a given growing season if deep soil water is available, which could result in watershed-scale water deficits. Woody encroachment also contributes to increased canopy interception of precipitation. Woody canopies typically intercept more rainfall than grass canopies, particularly when the woody community consists largely of coniferous species (mainly *Pinus* or *Juniperus* species in the United States) (Archer et al. 2017). Increased interception of rainfall further increases evapotranspiration in woody encroached areas compared to open grassy areas. Although impacts of woody encroachment on water yield vary

FIGURE 4.6 *Upper panel:* Lowveld savanna of Pilanesberg National Park, South Africa. The aboveground portions of the smaller trees (*Acacia* spp.) were completely killed by a recent fire and had begun to resprout. Note that tall trees have outgrown the "fire trap" and appear unaffected by the fire. *Lower panel:* A closer look at the resprouting *Acacia* spp. Note the dead branches with no leaves that were killed by the fire. The base of the tree was filled with resprouting shoots to recover from fire. The ability to resprout increases the persistence of woody species in highly disturbed environments such as grasslands and savannas. *Image credit*: Emily R. Wedel.

based on local climate and geomorphology (Huxman et al. 2005), woody encroachment generally results in an overall increase in evapotranspiration and decrease in groundwater recharge (Acharya et al. 2018). These impacts of vegetation change are typically most pronounced in mesic grasslands, where precipitation is high enough to recharge deeper soil water layers (Huxman et al. 2005).

In addition to increases in evapotranspiration, the proliferation of deeper, coarser root systems of woody species can impact water flow paths through the vadose zone (Zou et al. 2014; Acharya et al. 2018). Root systems have substantial impacts on the flow and retention of water in the soil profile (Cresswell et al. 1992; Scholl et al. 2014) as well as connectivity between water sources on a landscape. Coarse woody roots form large soil macropores more readily than finer grass roots, and these soil pores can alter flow paths and cause shifts in hydrologic partitioning in grassland systems by creating preferential flow paths deeper into the soil profile (Sullivan et al. 2019). These shifts have the potential to alter stream discharge and drainage through the vadose zone into groundwater, particularly during large rainfall events (Vero et al. 2018).

4.3.1.4 Specific Grassland Example – Juniper Encroachment in the Southern Great Plains, United States

Encroachment of Ashe juniper (*Juniperus ashei*) and eastern redcedar (*Juniperus virginiana*) in the southern Great Plains, United States, has resulted in substantial conversion of open grasslands and rangelands to juniper woodlands over the past 50–60 years (van Auken 2009). This transition is known to result in the loss of grassland small mammals and birds (including the now endangered lesser prairie chicken), reduced livestock production as the amount of quality forage declines and potentially alterations to local hydrologic cycles. Both juniper species have deeper rooting systems than the grass species they replace and can substantially alter surface soil conditions as hydrophobic litter layers are deposited (Wine et al. 2012). In addition to the loss of herbaceous forage for livestock, a major concern of land managers in the southern Great Plains has been the impact of juniper encroachment on streamflow and groundwater recharge (i.e., local water yield). However, the hydrologic impacts of juniper encroachment are not straightforward – they are heavily impacted by local climate conditions (especially annual precipitation) and geomorphology. Here, we will consider two contrasting consequences of juniper encroachment in different regions of the southern Great Plains.

In studies on southern Oklahoma, United States, in a region with relatively deep soil (1–2 m) underlain by shale and limestone bedrock, juniper encroachment has been linked to declines in runoff and streamflow. Surface runoff in grassland and rangeland systems in this region are typically dominated by infiltration excess overland flow – that is, when rainfall events result in saturated soil conditions, excess water that can no longer infiltrate contributes to surface runoff (Qiao et al. 2017). Juniper encroachment facilitates greater infiltration of water into deeper soil layers, resulting in less frequent soil-saturating rain events and lower rates of surface and subsurface runoff (Qiao et al. 2017). Rather than contributing to surface runoff and ultimately streamflow, infiltrating water instead contributes to recharge of deep soil water stores, which likely benefit deep-rooted junipers over more shallow-rooted grasses. Access to consistent deep soil water aids these juniper species in photosynthesizing year-round and in tolerating summer drought conditions.

In contrast, some studies on the Edwards Plateau in Texas have reported increased streamflow following encroachment by Ashe juniper (a species that is functionally equivalent to eastern red cedar (Qiao et al. 2017)). This region is characterized by a semiarid climate, shallow soils and highly permeable karst geology (Maclay 1995) where baseflow is the dominant contributor to streamflow (Wilcox and Huang 2010). In contrast to the above example, streamflow has actually increased through time as encroachment by Ashe juniper has progressed, and this change is not associated with a concurrent increase in precipitation (Wilcox and Huang 2010). The Edwards Plateau region experienced massive overgrazing by cattle from the late 1800s until roughly 1960, leading to overall degradation of existing rangelands (Wilcox and Huang 2010). Woody plants, particularly Ashe juniper, expanded after grazing pressures declined and increasing streamflow has been associated with improved infiltration of water to deeper portions of the soil. In this example, encroachment of Ashe juniper following the widespread land degradation in a karst geologic system led to an increase in streamflow rather than a decline. A separate study on juniper encroachment on the Edwards Plateau reported that removal of junipers did not result in an increase in groundwater recharge (Bazan et al. 2013). This example highlights the importance of land-use history – in combination with local climate, geomorphology and geology – in modulating the effects of woody encroachment on local water cycling.

As juniper encroachment becomes increasingly widespread in the southern Great Plains, understanding how this transition will impact runoff, streamflow and deep soil water recharge will be vital for land managers interested in maintaining rangeland forage quality and water resources in the future.

4.3.1.5 Specific Grassland Example – Subsection Invasive Species

In the United States alone, the negative impacts caused by invasive species accounts for nearly US$120 billion per year (Pimentel et al. 2005). Increased temperatures and precipitation variability associated with climate change puts native grassland communities at increased risk of invasion (Thomas et al. 2004). Exotic, invasive species are non-native and are typically accidentally introduced to a new grassland ecosystem by way of human activity. Invasive grassland species are able to survive and spread in the new ecosystem after being freed from the "restriction" of their native habitat. Native species are often replaced by invasive species as they become established and subsequently outcompete existing vegetation. For invasive grassland species that were intentionally introduced, the goal is commonly to increase forage quality for grazers or to reduce erosion. Unfortunately, these introductions typically result in altered species composition and declines in ecosystem productivity and biodiversity. Decreased diversity is often associated with loss of ecosystem resilience, or the ability of the ecosystem to tolerate disturbance such as fire or extreme climate events. In addition, replacement of diverse, native communities with monocultures of invasive species often leads to a decrease in productivity, in part due to alterations of fire dynamics and soil biogeochemistry.

Native plant species have evolved alongside the ecosystem they exist in – evolutionary pressures and competition with other native flora and fauna result in species that are well-adapted to their habitat. Changes to those conditions associated with human activity (e.g., nutrient additions, habitat fragmentation, increasing climate variability) are often much more rapid than the ability of species to change and adapt, resulting in opportunities for non-native species to establish and outcompete native species. Plant species that successfully invade grassland ecosystems typically contain innate characteristics that allow them to survive in a wide range of environmental and climatic conditions, allowing them to take advantage of novel and/or severe disturbances.

Particularly when coupled with changes in climate, human disturbance plays a key role in grassland invasion by exotic species. Grasslands with minimal human impacts historically have low rates of invasion. However, very few grasslands are free from anthropogenic impacts. Grasslands in Australia and North and South America are typically highly managed and have experienced high levels of fragmentation due to the increasing spread of agriculture and urbanization. These disturbances have led to increased invasibility, which can drastically alter species composition, productivity and ecosystem function, all of which have long-term ecological consequences (Gibson and Newman 2019).

4.3.1.6 Specific Grassland Example – Exotics in California Grasslands

California grasslands may be one of the best representations of an invaded ecosystem, as they have been converted from a diverse plant community with a large proportion dedicated to native bunchgrasses to an invaded landscape dominated by non-native Mediterranean annual grasses, which include, but are not limited to, grasses in the genera *Avena*, *Hordeum*, *Bromus* and *Schismus*. Such species have become dominant, as the climate in California is similar to the home ranges of these annual invasives. Interestingly, many of these California exotics do not dominate within their original range but serve as an early successional species (Jackson 1985). This is likely due to non-native species being better competitors for limiting resources and more tolerant to disturbances resulting from poor management strategies (high-intensity grazing) (HilleRisLambers et al. 2010).

The increase of human disturbances and rise of exotic annual species have resulted in the subsequent decline of native species. Not only is there a direct effect on other plant species, but there are also negative consequences such as increases in insects like aphids. Aphids are detrimental to plant health, as they feed on the carbohydrate-rich sap within phloem tissues, but they also serve as vectors for many plant viruses. Non-native annual grasslands have also been observed to alter soil nitrogen content and cycling that can result in long-term effects, mainly the deterrence of native perennial grasses reestablishment from the legacy of nitrogen-rich soils (Parker & Schimel 2010). While nitrogen serves as one of the most important macronutrients, it also serves to benefit fast-growing annual species. In this situation, the invasive annuals are able to outcompete slower growing native perennials. Not only are nitrogen cycles being altered by invasive species, but carbon dynamics have also been seen to increase in frequency and intensity. In contrast to the native perennial bunch grasses found in California, the invasive annuals have greater fuel loads and decreased fuel gaps which increase the probability and frequency of fires (Davies & Nafus 2013). The more frequent and intense fires moving across this region inhibit the growth of native species while simultaneously spurring the growth of invasives. All of the previously mentioned alterations to California grasslands were enabled and exacerbated by climate change, mainly human disturbances. Unfortunately for these grasslands, ecological dynamics now exist that reinforce the success of invasive species over native species and will require tremendous restoration and specific management practices to overcome.

4.3.1.7 Subsection Nutrient Deposition

Nitrogen availability varies among grassland ecosystems worldwide, tending to be lowest in semiarid temperate grasslands and in tropical grasslands with old soils

(McCulley et al. 2009, Risch et al. 2019). N availability is a key regulator of productivity across grasslands ecosystems worldwide (Stevens et al. 2015) and also impacts species richness and community characteristics (Midolo et al. 2019). Given that this critical resource is often limiting in many grassland ecosystems and grassland plant communities exhibit species-specific adaptations to this limitation, increased N availability has the potential to speed up local N cycling and alter ecosystem processes in addition to reordering species abundance to favor high N demanding species over those that adapted to low-N conditions.

One consequence of the green revolution is that synthetic fertilizer production now rivals (~100 Tg N yr[1]) amounts of naturally occurring N fixation at a global scale (Galloway et al. 2004). These synthetic fertilizers have been applied to improve agricultural yields to support our growing human population. While agronomic improvements of key crop species have reduced N requirements, and advances in application timing and amount have reduced N losses, a substantial portion of N is lost to runoff and leaching or is converted to gaseous forms and transported in the atmosphere to locations other than where originally applied. Rates of atmospheric N deposition are not uniform across regions. Regions with high acreages of commercial agriculture have higher rates of atmospheric transport and N deposition. Because many native grasslands were converted to row-crop agriculture, the remaining grassland ecosystems tend to be proximally associated with agricultural regions. As N accumulates in grasslands, it has the potential to eutrophy the local ecosystem, alter microbial dynamics and N cycling and reorder plant species populations (Johnson et al. 2008). These changes in N availability and a shift from a more N-limited system to N eutrophication represent a fundamental shift in grassland ecosystem dynamics and can stimulate negative outcomes via interactions with other global change drivers (e.g., rising air temperatures and more frequent drought).

4.4   CLIMATE ACTION SOLUTIONS

In this chapter, we have outlined the threats to grasslands posed by multiple climate change drivers, including elevated atmospheric $[CO_2]$, increased daytime and nighttime air temperatures, changes in precipitation patterns and intensity and altered nutrient deposition. The impacts of these drivers have taken over 150 years to develop and establish the trajectory of catastrophic global changes we are now experiencing. Our collective understanding of these threats and their impacts have developed over the past 30 years, as long-term data collection and observational ecological studies allowed for a documentation of change, and experimental studies facilitated the development of a mechanistic understanding of grassland responses to climate change. This research is composed of site-based and cross-site studies that have grown in complexity through time. Continuation of this research is necessary as effective and sustainable maintenance of grasslands requires better forecasting (both climatological and ecological) of grassland responses to climate change to develop effective climate solutions.

Developing a climate action plan is complex and difficult because the scope of climate change is large, the trajectory for recovery is long and the potential impacts of an individual or groups of individuals feel small. In the case of grasslands, solutions are especially overwhelming and pressing given the magnitude of the loss of grasslands due to agriculture, urbanization and degradation. Despite these difficulties in conserving grasslands, there is still a need for protecting and sustainably managing remaining grasslands to preserve their ecosystem services and biodiversity. The solution requires collective effort at the individual, community, national and ultimately global levels to influence policy, reduce the rate of climate change and conserve and restore grasslands (Figure 4.7). In this section, we explore options for climate actions, which if successful will result in meaningful long-term conservation and restoration of grassland ecosystems in the face of climate change.

4.4.1 Individual-Based Climate Actions That Promote Grassland Conservation in the Era of Global Change

At an individual level or within our local communities, it feels like an insurmountable challenge to make meaningful contributions to reducing global CO_2 emissions when governments have the greatest influence through net-zero energy policies that limit the choices of industries and individuals to reduce carbon emissions. While mitigating the trajectory of climate change requires fundamental policy changes at the national and global scale, there are several individual- and community choice-based steps that are impactful with collective action: (1) Exercise your consumer choices that prioritize supporting businesses that invest in renewable/net-zero C emission energy policies. While small individually, collective consumerism can enact progressive change. (2) Develop a deeper understanding of how increased CO_2 emissions impact our climate system, and why changes in temperature and precipitation negatively impact grassland ecosystems. The details of climate science can be confusing and has led to inaction for many individuals and nations. Spending time to learn how this change has happened, and why the threats are so dire will allow you to factually inform family and friends about this process. Thorough, simple and fact-based education is the key to dispelling misinformation and raising awareness that leads to action. (3) Support international conservation programs whose mission is to protect natural landscapes and recover degraded areas back into natural habitats. Organizations like The Nature Conservancy (among others) have a mission

Global	Regional	Individual
• Reduce GHG emissions	• Policy change	• Consumer change
• Sustainable agriculture	• Sustainable land management	• Education / outreach
• No afforestation	• Restoration of native grasslands	• Support organizations

FIGURE 4.7 Climate actions for impactful long-term grassland conservation and restoration are possible at multiple levels. Section 4.4 outlines achievable actions possible at individual, regional and global levels.

to protect grassland ecosystems worldwide, and have the organization structure required to impact local, regional and national policies. Given that these organizations are nonprofit, they rely upon the support of the public to maintain their mission and continue to enact impactful climate actions.

4.4.2 CONSERVING AND RESTORING LOCAL GRASSLANDS

Protecting the grassland habitat that remains is the most impactful activity that we can all engage in. As noted throughout this chapter, grassland ecosystems throughout the world have been greatly impacted and transformed into agriculture, urban environments and degraded environments. The grassland habitat that remains – whether in natural environments or as fragments within urban environments or agricultural corridors – have intrinsic value. Mosaics of remnant grasslands provide key refugia for plant and animal species, resulting in increases in plant and animal diversity, as well as beneficial local ecosystem services like reduced runoff/erosion, increased C-storage belowground (in grassy plant roots), pollination services, reduced disease threats and habitat for migratory bird species. Given that functional grassland ecosystems take decades to centuries to establish (Baer et al. 2020), protection of existing grasslands should always be prioritized as our most important conservation action.

In the United States and likely in many locations throughout the world, local grassland societies have been organized. These societies typically have dual missions to protect specific locations, or regional grasslands, to provide educational outreach and to provide recreational opportunities for the public to engage with grassland ecosystems. Local grassland societies are powerful advocates for protecting remnant grasslands and maintaining functional grassland mosaics across the landscape. Supporting these local grassland societies with financial contributions or with your personal service and advocacy is one of the most meaningful climate actions available. At a larger scale, national and international conservation organizations commonly engage with landowners to establish Land Trusts. Land Trusts are established as part of estate planning and specify extended periods of time whereby natural lands remain undeveloped for agriculture or for urbanization. These activities ensure that grasslands will continue to provide functional ecosystem services across generations.

Conservation of intact ("natural") grasslands is definitively the most important climate action available for grassland ecosystems. As noted previously in this chapter, woody plant encroachment in grassland ecosystems is driven by a combination of global climate drivers and changes in local land-use drivers. This conversion of grassland ecosystems from grassy-dominated to woody-dominated results in a fundamental shift in ecosystem properties and services, and typically results in biodiversity losses (Ratajczak et al. 2012). Because the transition from grassland to shrubland/woodland-dominated represents an alternative stable state (Ratajczak et al. 2017), hysteresis in the system often slows recovery of the original grassland ecosystem (Collins et al. 2021) or requires intensive management options (Nippert et al. 2021). Specifically, because the transition from a shrub-dominated back to a grassy-dominated ecosystem is so challenging, the best course of action involves management plans that restrict woody plant expansion in the first place. Typically, this includes sound management policies that prioritize regular burning on the landscape (Twidwell et al. 2013). For grassland locations where prevention of woody encroachment is no longer possible, prescriptive policies to physically remove woody plants (i.e., brush-cutting and tree removal) are often the only option. These campaigns are often laborious and expensive. However, many states agencies have funding available for woody removal as part of broader conservation goals. Limiting the spread of woody encroachment or implementing tree removal are a key first step for ecological restoration of grassland ecosystems.

Given the highly impacted nature of grassland ecosystems, ecological restoration is widely utilized to recover diversity and ecological functioning in previously degraded grasslands. As already mentioned, restoration can occur following woody encroachment, in locations previously converted to agriculture, or following removal of invasive species. Engagement with grassland restoration provides many opportunities to promote meaningful climate actions. Perhaps most importantly, the process of reintroducing native plant species (and native genotypes) into the environment has carryover positive impacts that initiate the process of restoring soil fertility, growing healthy root systems that store carbon in the soil, hydrological benefits that include reduced erosion and leaching as well as increased infiltration pathways (leading to vadose-zone soil moisture recharge). Grassland restoration also provides habitat for vertebrates and invertebrates species. For example, using diverse assemblages of local plant species provides increased niche space for specialized invertebrates, increased pollination services and an ecological platform that facilitates greater abundance and diversity of vertebrate species. The process of grassland restoration doesn't only apply to large acreage locations. There can be large benefits from establishing small grassland communities in urban environments. The simple act of replacing turf with mixes of native species reduces eutrophication and runoff (from lawn fertilizers), helps conserve water since native species are often more water-use efficient than turfgrass cultivars and provides ecosystem services like pollination and habitat for birds, small mammals and insects.

4.4.3 ENGAGEMENT IN COMMUNITY EDUCATION

Developing a long-term culture of appreciation for the natural world often begins with effective K–12 educational programs that seek to teach children about nature. Teaching and developing an appreciation for grassland ecology and grassland ecosystems at a young age is key to developing an informed population that values these lands for their inherent services beyond a commodity-based value system only. Starting with the youngest age groups, science educators are able to instill wonderment about grassland ecosystems, the species they harbor and the ecosystem services they provide. With older children, these basic concepts can be supplemented with science-based inquiry that explains the threat of climate change to grasslands and the corresponding impacts on local plant, animal and human communities. Building scientific concepts through time facilitates educational scaffolding providing both breadth and depth of scientific understanding in children and develops a population that both values *and* understands the natural world. If you are a local educator, reach out to local grassland societies or conservation groups for suggestions on incorporating educational materials, including field trips, within your curriculum.

Don't forget that many online curriculums and programs exist! If you are a parent, encourage your children's teachers and administrators to engage in grassland ecology education and science-based discussions of climate change drivers, impacts and consequences. If you are engaged citizen, encourage your local city councils and commissioners to support science-based and nature-based educational programs within your communities. Often, increased science-learning and science-based educational opportunities can happen with a few field trips per year, online engagement with outreach coordinators from grassland ecosystems and typically do not require large increases in financial expenditure. Perhaps more than any other climate action suggested, developing an appreciation and understanding of climate change impacts and climate action for grassland ecosystems among K–12 students has the most significant potential for long-term substantial climate actions.

4.4.4 GLOBAL ACTIONS

There is no viable scientific explanation for these climate changes that doesn't include increased anthropogenic fossil fuel emissions as the primary driver of change. For this reason, there is no viable solution to the problem that doesn't include a large-scale reduction in fossil fuel emissions and the resultant CO_2 emissions of this energy-conversion process. Any "solution" that doesn't include drastic reductions in CO_2 is untenable. For this simple reason, impactful climate action for the long-term conservation of natural ecosystems like grasslands must include a detailed transition away from C-based energy sources toward zero net-emission energy sources.

Long-term observational and experimental research has increased our understanding of the impacts of climate change on grasslands and has led to an appreciation of the mechanisms that maintain grassland stability in response to disturbance. The biodiversity and stability of natural grasslands have become a model for sustainable agriculture, which aims to maintain economic viability while improving social equity and protecting environmental health and natural resources. These efforts, inspired by natural systems, include increasing biodiversity to help mitigate pest outbreaks and agrosystem resilience and patch-burn grazing that increases ecosystem biodiversity and improves habitat for wildlife. Transformative agricultural organizations like The Land Institute (Salina, KS, USA) use native grass species to develop deep-rooted perennial crops that protect soil health and increase C sequestration. The goal of perennial agriculture is to create productive, profitable agricultural systems that mimic the benefits of carbon and water cycling in grasslands. Thus, long-term grassland and climate research is directly related to the development of sustainable agriculture, which is becoming increasingly pressing in the face of climate change and a growing human population.

4.4.4.1 Case against Global Tree-Planting Campaigns in Grasslands

"Let Us Plant the Right Trees, in the Right Place, for the Right Reasons"

– William Bond.

It's tempting to look for actionable items that have been effective for climate change mitigation in other ecosystems, and then apply them to grasslands. One climate action that is relatively easy to implement, has widespread public enthusiasm and can be performed over large regions has been the sponsorship of tree-planting campaigns in deforested regions. Increased tree planting has the potential to sequester increased amounts of atmospheric CO_2 in woody biomass and alter radiative forcing and energy exchange through the development of complex tree canopies. Unfortunately, these "good ideas" to plant trees are often applied to afforested regions (Bond 2019), which typically include grasslands. A common misconception by the public is that the "good" of planting trees as a feasible climate action outweighs any inherent ecosystem services and innate climate buffering provided by native grassland communities. This is not the case, as grassland ecosystems are not a midpoint along a continuum from a degraded state to a forested state (Veldman et al. 2015). Planting trees in locations where trees have not previously existed in high abundance risks major biodiversity losses, changes in biogeochemical cycling, altered fire risks and increased water fluxes.

Perhaps a more appropriate climate action to offset the impacts of global changes is widespread perennial grass-planting campaigns. As we've already discussed, native grassland species provide a wide array of ecosystem services. Grassland ecosystems are often quite comparable in carbon sequestration to forests, especially when species with dense and deep root systems establish. Global climate change has increased the frequency and intensity of droughts for many locations worldwide. Grassy species are more resilient to drought, and their recovery from drought is often more robust compared to woody plant species (Choat et al. 2018). Widespread tree-planting campaigns intended for locations that experience natural and/or climate-change-driven drought may actually accelerate carbon losses. When forests are hit with drought and die, these locations are then more susceptible to fire, and fire-based C emissions (Dass et al. 2018). For these reasons, when the risks of fire and climate uncertainty (i.e., drought) are considered, the benefits of planting trees in grassland regions as a climate action seem clearly misguided.

4.4.5 CONCLUSION

Grasslands provide tremendous ecological and societal value worldwide. These ecosystems are key regulators of global processes like carbon, nutrient and water cycling as well as serving as key reservoirs of biological species diversity. As we've specified in this chapter, the threats of climate change are acutely present in grasslands. These impacts may vary across continents and grassland types, but there is a similar negative consequence of loss of function and a reduction in diversity that is felt worldwide. For these reasons, climate actions must be undertaken immediately. As outlined in this section, climate action plans can be prescribed at multiple scales (Figure 4.7). We are each able to begin making an impact within our local communities. These climate actions include our choices as consumers as well as supporting local grassroots organizations to protect and restore grasslands. The actions we take as individuals impact regional policies and education and outreach to promote sustainable grassland management with regulations to minimize further degradation and reward public and private entities involved in grassland restoration. These regional changes will ultimately translate to large-scale (global) climate actions that reduce greenhouse gas emissions, prioritize grasslands and avoid false solutions like afforestation of grasslands. Working together at both local and national scales, we can help implement actions and policies that slow the negative impacts of climate change, and through time recover these ecosystems. Doing so, we will be protecting an ecosystem that provides beauty, key human services and an inherent sense of place to billions of people worldwide.

LITERATURE CITED

Acharya, B. S., Hao, Y., Ochsner, T. E., & Zou, C. B. (2017). Woody plant encroachment alters soil hydrological properties and reduces downward flux of water in tallgrass prairie. *Plant and Soil, 414*(1–2), 379–391.

Acharya, B. S., Kharel, G., Zou, C. B., Wilcox, B. P., & Halihan, T. (2018). Woody plant encroachment impacts on groundwater recharge: A review. *Water, 10*(10), 1466.

Ainsworth, E. A., & Ort, D. R. (2010). How do we improve crop production in a warming world? *Plant Physiology, 154*(2), 526–530.

Allred, B. W., Scasta, J. D., Hovick, T. J., Fuhlendorf, S. D., & Hamilton, R. G. (2014). Spatial heterogeneity livestock productivity in a changing climate. *Agriculture, Ecosystem and Environment, 193*, 37–41.

Archer, S. R., Andersen, E. M., Predick, K. I., Schwinning, S., Steidl, R. J., & Woods, S. R. (2017). Woody plant encroachment: Causes and consequences. In D. D. Briske (Ed.), *Rangeland systems: Processes, management and challenges* (pp. 25–84). Cham, Switzerland: Springer.

Baer, S. G., Adams, T., Scott, D. A., Blair, J. M., & Collins, S. L. (2020). Soil heterogeneity increases plant diversity after 20 years of manipulation during grassland restoration. *Ecological Applications, 30*(1), e02014. doi: 10.1002/eap.2014.

Bazan, R. A., Wilcox, B. P., Munster, C., & Gary, M. (2013). Removing woody vegetation has little effect on conduit flow recharge. *Ecohydrology, 6*(3), 435–443.

Bestelmeyer, B. T., Ellison, A. M., Fraser, W. R., Gorman, K. B., Holbrook, S. J., Laney, C. M., … Sharma, S. (2011). Analysis of abrupt transitions in ecological systems. *Ecosphere, 2*(12), 1–26.

Blair, J., Nippert, J. B., & Briggs, J. M. (2014). Grassland ecology. In R. K. Monson (Ed.), *The plant sciences – Ecology & the environment* (pp. 389–423). Berlin: Springer-Verlag Berlin Heidelberg.

Bond, W. J. (2019). *Open ecosystems: Ecology and evolution beyond the forest edge.* Oxford, UK: Oxford University Press.

Briggs, J. M., & Knapp, A. K. (1995). Interannual variability in primary production in tallgrass prairie: Climate, soil moisture, topographic position, and fire as determinants of aboveground biomass. *American Journal of Botany, 82*(8), 1024–1030. doi: 10.1002/j.1537-2197.1995.tb11567.x.

Briggs, J. M., Knapp, A. K., Blair, J. M., Heisler, J. L., Hoch, G. A., Lett, M. S., & McCarron, J. K. (2005). An ecosystem in transition: Causes and consequences of the conversion of mesic Grassland to Shrubland. *BioScience, 55*(3), 243–254. doi: 10.1641/0006-3568(2005)055[0243:AEITCA]2.0.CO;2.

Brooks, M. L., D'antonio, C. M., Richardson, D. M., Grace, J. B., Keeley, J. E., DiTomaso, J. M., … Pyke, D. (2004). Effects of invasive alien plants on fire regimes. *BioScience, 54*(7), 677–688.

Case, M. F., & Staver, A. C. (2017). Fire prevents woody encroachment only at higher-than-historical frequencies in a South African savanna. *Journal of Applied Ecology, 54*(3), 955–962.

Choat, B., Brodribb, T. J., Brodersen, C. R., Duursma, R. A., López, R., & Medlyn, B. E. (2018). Triggers of tree mortality under drought. *Nature, 558*(7711), 531–539.

Collins, S. L., Nippert, J. B., Blair, J. M., Briggs, J. M., Blackmore, P., & Ratajczak, Z. (2021). Fire frequency, state change and hysteresis in tallgrass prairie. *Ecology Letters, 24*(4), 636–647.

Cresswell, H. P., Smiles, D. E., & Williams, J. (1992). Soil structure, soil hydraulic properties and the soil water balance. *Australian Journal of Soil Research, 30*(3), 265–283.

Dass, P., Houlton, B. Z., Wang, Y., & Warlind, D. (2018). Grasslands may be more reliable carbon sinks than forests in California. *Environmental Research Letters, 13*(7), 1–8.

Davies, K. W., & Nafus, A. M. (2013). Exotic annual grass invasion alters fuel amounts, continuity and moisture content. *International Journal of Wildland Fire, 22*(3), 353–358. doi: 10.1071/WF11161.

Dijkstra, F. A., Blumenthal, D., Morgan, J. A., LeCain, D. R., & Follett, R. F. (2010). Elevated CO2 effects on semi-arid grassland plants in relation to water availability and competition. *Functional Ecology, 24*(5), 1152–1161. doi: 10.1111/j.1365-2435.2010.01717.x.

Dixon, A. P., Faber-Langendoen, D., Josse, C., Morrison, J., & Loucks, C. J. (2014). Distribution mapping of world grassland types. *Journal of Biogeography, 41*(11), 2003–2019.

Drewa, P. B., & Havstad, K. M. (2001). Effects of fire, grazing, and the presence of shrubs on Chihuahuan Desert grasslands. *Journal of Arid Environments, 48*(4), 429–443.

Fay, P. A., Blair, J. M., Smith, M. D., Nippert, J. B., Carlisle, J. D., & Knapp, A. K. (2011). Relative effects of precipitation variability and warming on tallgrass prairie ecosystem function. *Biogeosciences, 8*(10), 3053–3068.

Fay, P. A., Carlisle, J. D., Knapp, A. K., Blair, J. M., & Collins, S. L. (2003). Productivity responses to altered rainfall patterns in a C4-dominated grassland. *Oecologia, 137*(2), 245–251.

Felton, A. J., Knapp, A. K., & Smith, M. D. (2019). Carbon exchange responses of a mesic grassland to an extreme gradient of precipitation. *Oecologia, 189*(3), 565–576.

Galloway, J. N., Dentener, F. J., Capone, D. G., Boyer, E. W., Howarth, R. W., Seitzinger, S. P., … Vöosmarty, C. J. (2004). Nitrogen cycles: Past, present, and Future. *Biogeochemistry, 70*(2), 153–226.

Gibson, D. J. (2009). *Grasses and grassland ecology*. New York: Oxford University Press.

Gibson, D. J., & Newman, J. A. (2019). *Grasslands and climate change*. Cambridge, UK: Cambridge University Press. doi: 10.1017/9781108163941.

Giorgi, F., Raffaele, F., & Coppola, E. (2019). The response of precipitation characteristics to global warming from climate projections. *Earth System Dynamics, 10*(1), 73–89.

Gray, E. F., & Bond, W. J. (2013). Will woody plant encroachment impact the visitor experience and economy of conservation areas? *Koedoe, 55*(1), 1–9.

Hartnett, D. C., Hickman, K. R., & Walter, L. E. F. (1996). Effects of Bison Grazing, fire, and topography on floristic diversity in tallgrass Prairie. *Journal of Range Management, 49*(5), 413–420.

Hayden, B. P. (1998). Regional climate and the distribution of tallgrass prairie. In A. K. Knapp, J. M. Briggs, D. C. Hartnett & S. L. Collins (Eds.), *Grassland dynamics* (pp. 19–34). New York: Oxford Press.

Heisler-White, J., Knapp, A. K., & Kelly, E. F. (2008). Increasing precipitation event size increases aboveground net primary productivity in a semi-arid grassland. *Oecologia, 158*(1), 129–140.

Heisler-White, J. L., Blair, J. M., Kelly, E. F., Harmony, K., & Knapp, A. K. (2009). Contingent productivity responses to more extreme rainfall regimes across a grassland biome. *Global Change Biology, 15*(12), 2894–2904. doi: 10.1111/j.1365-2486.2009.01961.x.

HilleRisLambers, J., Yelenik, S. G., Colman, B. P., & Levine, J. M. (2010). California annual grass invaders: The drivers or passengers of change? *Journal of Ecology, 98*(5), 1147–1156. doi: 10.1111/j.1365-2745.2010.01706.x.

Hoover, D. L., Knapp, A. K., & Smith, M. D. (2014). Resistance and resilience of a grassland ecosystem to climate extremes. *Ecology, 95*(9), 2646–2656. doi: 10.1890/13-2186.1.

Huxman, T. E., Wilcox, B. P., Breshears, D. D., Scott, R. L., Snyder, K. A., Small, E. E., … Jackson, R. B. (2005). Ecohydrological implications of woody plant encroachment. *Ecology, 86*(2), 308–319.

IPCC. (2007). Observations and atmospheric climate change. In: *Climate change 2007: The physical science basis. Contribution of working group I to the Fourth Assessment report of the intergovernmental panel on climate change.* Cambridge: Cambridge University Press.

IPCC. (2019). *Climate change and land: An IPCC special report on climate change, desertification, land degradation, sustainable land management, food security, and greenhouse gas fluxes in terrestrial ecosystems. Contributions of working group to the assessment report of the intergovernmental panel on climate change.* Cambridge: Cambridge University Press.

Jackson, L. E. (1985). Ecological origins of California's Mediterranean grasses. *Journal of Biogeography, 12*(4), 349–361.

Johnson, N. C., Rowland, D. L., Corkidi, L., & Allen, E. B. (2008). Plant winners and losers during grassland N-eutrophication differ in biomass allocation and mycorrhizas. *Ecology, 89*(10), 2868–2878. doi: 10.1890/07-1394.1.

Knapp, A. K., Briggs, J. M., Hartnett, D. C., & Collins, S. L. (1998). *Grassland dynamics: Long-term ecological research in tallgrass Prairie.* New York: Oxford University Press.

Knapp, A. K., Chen, A., Griffin-Nolan, R. J., Baur, L. E., Carroll, C. J. W., Gray, J. E., … Smith, M. D. (2020). Resolving the Dust Bowl paradox of grassland responses to extreme drought. *Proceedings of the National Academy of Sciences, 117*(36), 22249–22255.

Knapp, A., Ciais, P., & Smith, M. (2016). Reconciling inconsistencies in precipitation-productivity relationships: Implications for climate change. *New Phytologist, 214*(1), 41–47.

Knapp, A., Fay, P., Blair, J., Collins, S. L., Smith, M. D., Carlisle, J. D., … McCarron, J. K. (2002). Rainfall variability, carbon cycling, and plant species diversity in a mesic grassland. *Science, 298*(5601), 2202–2205. doi: 10.1126/science.1076347.

Körner, C. (2006). Plant CO_2 responses: An issue of definition, time, and resource supply. *New Phytologist, 172*(3), 393–411.

Krauss, J., Bommarco, R., Guardiola, M., Heikkinen, R. K., Helm, A., Kuussaari, M. (2010). Habitat fragmentation causes immediate and time-delayed biodiversity loss at different trophic levels. *Ecology Letters, 13*(5), 597–605.

Kulmatiski, A., & Beard, K. H. (2013). Woody plant encroachment facilitated by increased precipitation intensity. *Nature Climate Change, 3*(9), 833–837.

Maclay, R. W. (1995). Geology and hydrology of the Edwards aquifer in the San Antonio area, Texas. *Water-resources investigations report 95-4186.* U.S. Geological Survey, Austin, TX.

McCulley, R. L., Burke, I. C., & Lauenroth, W. K. (2009). Conservation of nitrogen increases with precipitation across a major grassland gradient in the Central Great Plains of North America. *Oecologia, 159*(3), 571–581.

Midolo, G., Alkemade, R., Schipper, A. M., Benítez-López, A., Perring, M. P., & De Vries, W. (2019). Impacts of nitrogen addition on plant species richness and abundance: A global meta-analysis. *Global Ecology and Biogeography, 28*(3), 398–413.

Miller, J. E., Damschen, E. I., Ratajczak, Z., & Özdoğan, M. (2017). Holding the line: Three decades of prescribed fires halt but do not reverse woody encroachment in grasslands. *Landscape Ecology, 32*(12), 2297–2310.

Nippert, J. B., Knapp, A. K., & Briggs, J. M. (2006). Intra-annual rainfall variability and grassland productivity: Can the past predict the future? *Plant Ecology, 184*(1), 65–74. doi: 10.1007/s11258-005-9052-9.

Nippert, J. B., Telleria, L., Blackmore, P., Taylor, J. H., & O'Connor, R. C. (2021). Is a prescribed fire sufficient to slow the spread of woody plants in an infrequently burned grassland? A case study in tallgrass prairie. *Rangeland Ecology and Management, 78*, 79–89.

O'Keefe, K., Bell, D. M., McCulloh, K. A., & Nippert, J. B. (2020). Bridging the flux gap: Sap flow measurements reveal species-specific patterns of water use in a tallgrass Prairie. *Journal of Geophysical Research: Biogeosciences, 125*(2), e2019JG005446.

Parker, S. S., & Schimel, J. P. (2010). Invasive grasses increase nitrogen availability in California grassland soils. *Invasive Plant Science and Management, 3*(1), 40–47. doi: 10.1614/ipsm-09-046.1.

Pearcy, R. W., & Ehleringer, J. (1984). Comparative ecophysiology of C3 and C4 plants. *Plant, Cell and Environment, 7*(1), 1–13.

Pendergrass, A. G., Knutti, R., Lehner, F., Deser, C., & Sanderson, B. M. (2017). Precipitation variability increases in a warmer climate. *Scientific Reports, 7*(1), 17966.

Pimentel, D., Zuniga, R., & Morrison, D. (2005). Update on the environmental and economic costs associated with alien-invasive species in the United States. *Ecological Economics, 52*(3), 273–288. doi: 10.1016/j.ecolecon.2004.10.002.

Qiao, L., Zou, C. B., Stebler, E., & Will, R. E. (2017). Woody plant encroachment reduces annual runoff and shifts runoff mechanisms in the tallgrass prairie, USA. *Water Resources Research, 53*(6), 4838–4849.

Ramankutty, N., Evan, A. T., Monfreda, C., & Foley, J. A. (2008). Farming the planet: 1. Geographic distribution of global agricultural lands in the year. *Global Biogeochemistry Cycles, 22*(1), 1–19. doi: 10.1029/2007GB002952.

Ratajczak, Z., D'Odorico, P., Collins, S. L., Bestelmeyer, B. T., Isbell, F. I., & Nippert, J. B. (2017). The interactive effects of press/pulse intensity and duration on regime shifts at multiple scales. *Ecological Monographs, 87*(2), 198–218. doi: 10.1002/ecm.124.

Ratajczak, Z., Nippert, J. B., & Collins, S. L. (2012). Woody encroachment decreases diversity across North American grasslands and savannas. *Ecology, 93*(4), 697–703. doi: 10.1890/11-1199.1.

Ratajczak, Z., Nippert, J. B., & Ocheltree, T. (2014). Abrupt transition of mesic grassland to shrubland: Evidence for thresholds, alternative attractors, and regime shifts. *Ecology, 95*(9), 2633–2645.

Reich, P. B., Sendall, K. M., Stefanski, A., Rich, R. L., Hobbie, S. E., & Montgomery, R. A. (2018). Effects of climate warming on photosynthesis in boreal tree species depend on soil moisture. *Nature, 562*(7726), 263. doi: 10.1038/s41586-018-0582-4.

Risch, A. C., Zimmermann, S., Ochoa-Hueso, R., Schütz, M., Frey, B., Firn, J. L., … Moser, B. (2019). Soil net nitrogen mineralisation acro1ss global grasslands. *Nature Communications, 10*(1), 4981. doi: 10.1038/s41467-019-12948-2.

Sage, R. F., & Kubien, D. S. (2007). The temperature response of C3 and C4 photosynthesis. *Plant, Cell and Environment, 30*(9), 1086–1106.

Sankaran, M., Ratnam, J., & Hanan, N. (2008). Woody cover in African savannas: The role of resources, fire and herbivory. *Global Ecology and Biogeography, 17*(2), 236–245.

Scholl, P., Leitner, D., Kammerer, G., Loiskandl, W., Kaul, H. P., & Bodner, G. (2014). Root induced changes of effective 1D hydraulic properties in a soil column. *Plant and Soil, 381*(1–2), 193–213.

Smit, I. J., & Prins, H. H. (2015). Predicting the effects of woody encroachment on mammal communities, grazing biomass and fire frequency in African Savannas. *PLOS ONE, 10*(9), e0137857.

Smith, M. D. (2011). The ecological role of climate extremes: Current understanding and future prospects. *Journal of Ecology, 99*(3), 651–655. doi: 10.1111/j.1365-2745.2011.01833.x.

Staver, A. C., Archibald, S., & Levin, S. A. (2011). The global extent and determinants of savanna and forest as alternative biome states. *Science, 334*(6053), 230–232.

Stevens, C. J., Lind, E. M., Hautier, Y., Harpole, W. S., Borer, E. T., Hobbie, S., ... Wragg, P. D. (2015). Anthropogenic nitrogen deposition predicts local grassland primary production worldwide. *Ecology, 96*(6), 1459–1465.

Stevens, N., Erasmus, B. F. N., Archibald, S., & Bond, W. J. (2016). Woody encroachment over 70 years in South African savannahs: Overgrazing, global change or extinction aftershock? *Philosophical Transactions of the Royal Society of London Series B, 371*, 20150437.

Strömberg, C. A. (2011). Evolution of grasses and grassland ecosystems. *Annual Review of Earth and Planetary Sciences, 39*(1), 517–544.

Sullivan, P. L., Stops, M. W., Macpherson, G. L., Li, L., Hirmas, D. R., & Dodds, W. K. (2019). How landscape heterogeneity governs stream water concentration-discharge behavior in carbonate terrains (Konza Prairie, USA). *Chemical Geology, 527*, 118989. doi: 10.1016/j.chemgeo.2018.12.002.

Thomas, C. D., Cameron, A., Green, R. E., Bakkenes, M., Beaumont, L. J., Collingham, Y. C., ... Williams, S. E. (2004). Extinction risk from climate change. *Nature, 427*(6970), 145–148. doi: 10.1038/nature02121.

Tilman, D., Balzer, C., Hill, J., & Befort, B. L. (2011). Global food demand and the sustainable intensification of agriculture. *Proceedings of the National Academy of Sciences, 108*(50), 20260–20264.

Tjoelker, M. G., Oleksyn, J., & Reich, P. B. (2001). Modeling respiration of vegetation: Evidence for a general temperature-dependent Q10. *Global Change Biology, 7*(2), 223–230.

Twidwell, D., Rogers, W. E., Fuhlendorf, S. D., Wonkka, C. L., Engle, D. M., Weir, J. R., ... Taylor, C. A. (2013). The rising Great Plains fire campaign: Citizens' response to woody plant encroachment. *Frontiers in Ecology and the Environment, 11*(s1), e64–e71. doi: 10.1890/130015.

U.S. Global Change Research Program. (2018). *Impacts, risks, and adaptation in the United States: Fourth national climate assessment*, volume 2. D. R. Reidmiller et al. (Eds.). Washington, DC: U.S. Global Change Research Program.

Van Auken, O. W. (2009). Causes and consequences of woody plant encroachment into western North American grasslands. *Journal of Environmental Management, 90*(10), 2931–2942.

Veldman, J. W., Buisson, E., Durigan, G., Fernandes, G. W., Le Stradic, S., Mahy, G., ... Bond, W. J. (2015). Toward an old-growth concept for grasslands, savannas, and woodlands. *Frontiers in Ecology and the Environment, 13*(3), 154–162. doi: 10.1890/140270.

Venter, Z. S., Cramer, M. D., & Hawkins, H. J. (2018). Drivers of woody plant encroachment over Africa. *Nature Communications, 9*(1), 1–7.

Vero, S. E., Macpherson, G. L., Sullivan, P. L., Brookfield, A. E., Nippert, J. B., Kirk, M. F., ... Kempton, P. (2018). Developing a conceptual framework of landscape and hydrology on tallgrass prairie: A critical zone approach. *Vadose Zone Journal, 17*(1), 1–11.

Wilcox, B. P., & Huang, Y. (2010). Woody plant encroachment paradox: Rivers rebound as degraded grasslands convert to woodlands. *Geophysical Research Letters, 37*(7), L07402.

Wilcox, K. R., Shi, Z., Gherardi, L. A., Lemoine, N. P., Koerner, S. E., Hoover, D. L., ... Luo, Y. (2017). Asymmetric responses of primary productivity to precipitation extremes: A synthesis of grassland precipitation manipulation experiments. *Global Change Biology, 23*(10), 4376–4385. doi: 10.1111/gcb.13706.

Wine, M. L., Ochsner, T. E., Sutradhar, A., & Pepin, R. (2012). Effects of eastern redcedar encroachment on soil hydraulic properties along Oklahoma's grassland-forest ecotone. *Hydrological Processes, 26*(11), 1720–1728.

Zou, C. B., Turton, D. J., Will, R. E., Enge, D. M., & Fuhlendorf, S. D. (2014). Alteration of hydrological processes and streamflow with juniper (*Juniperus virginiana*) encroachment in a mesic grassland catchment. *Hydrological Processes, 28*(26), 6173–6182.

5 Securing Urban Water Systems in a Changing Climate in the San Francisco Bay Area, California

Sasha Harris-Lovett, and Richard G. Luthy

CONTENTS

5.1 INTRODUCTION

April 1 marks the official annual spring snow survey in California's Sierra Nevada Mountains, a time of year that typically represents the highest water content before the spring sunshine causes accumulated snowdrifts to melt. This year in 2021, snowpack across the state hovered at about half of normal (California Department

DOI: 10.1201/9781003048701-5

of Water Resources 2021b). A spokesperson from the California Department of Water Resources told reporters it looked like "one of the 10 worst snowpacks in California history" (Arcuni 2020). The year 2021 is also the second dry year in a row, which means some of the state's reservoirs started the year well below their capacity (California Department of Water Resources 2021b). These measurements have implications that ripple across the entire state, because many of California's urban areas and agricultural operations rely upon snowmelt for their water supply.

California is no stranger to drought, and more is on the horizon. Warming temperatures due to anthropogenic climate change will increase the risk of longer, drier droughts in the state (Diffenbaugh et al. 2015). It is becoming increasingly clear that current systems for supplying drinking water to Californians are vulnerable and must be adapted to support the well-being of many cities, including that of the metropolitan San Francisco Bay Area. While technologies and strategies to adapt urban water systems to climate change exist and more are in development, numerous challenges impede their widespread adoption. Regional collaborative networks and bridges among institutional silos and communities are imperative for effectively and equitably adapting urban water systems for a changing climate. Developing these new systems at a regional scale holds great promise for respecting unique cultural needs and geographic realities while also taking advantages of economies of scale and regional-level resources.

By looking closely at the case study of the San Francisco Bay Area, California, this chapter illuminates some of the challenges climate change poses to urban water supplies in arid and Mediterranean climates; describes a few of the potential solutions; and clarifies the key next steps for advancing more resilient urban water systems.

5.2 A BRIEF HISTORY OF THE BAY AREA'S WATER SUPPLIES

The Bay Area, like much of California, has long relied on imported water to meet the demands of a thirsty, growing population. In 2021, this nine-county region was home to more than seven million people (Association of Bay Area Governments 2021). The Bay Area itself has a Mediterranean climate and is approximately 200 miles away from the Sierra Nevada Mountains (the Sierra). Snowmelt from the Sierra is diverted to urban water purveyors and comprises about two-thirds of the Bay Area's water supply (Ackerly et al. 2018).

5.2.1 Importing Water from Afar

Historically, engineers have considered California's water to be a commodity that can and should be moved to where it can provide the most economic impact. The San Francisco Bay Area has long exemplified this mindset. In the early 1900s, one-third of San Francisco's population didn't have running water and had to get it delivered in barrels and wagons (Martin 1914). San Franciscans looked to the snowy Sierras and their snowmelt-fed rivers for a solution. In 1901, the San Francisco city engineer described the advantages of capturing and diverting the Tuolumne River to

San Francisco via a proposed dam at the Hetch Hetchy valley. He described the Tuolumne River as having

> absolute purity by reason of the uninhabitable character of the entire watershed tribu-tary to the reservoirs, and ... abundance far beyond possible future demands for all purposes. ... Considering the partial and rapid rate of pollution to which all other sources may in the future be subjected, particularly near-by sources, the Tuolumne River is far superior to any other.
>
> **(Martin 1914)**

Urban residents felt entitled to the precipitation that fell in other parts of the state and had little regards for ecological or Indigenous uses for water.

The City of San Francisco then proceeded to pass a bond measure to buy the privately owned land around the Hetch Hetchy valley, along with the water rights and timber rights in the surrounding area (Martin 1914). There was just one catch: the proposed dam site was within the borders of Yosemite National Park. San Francisco threw its support behind the Raker Bill, a proposal to allow the City of San Francisco to dam the Hetch Hetchy valley and divert the Tuolumne River water to the city. The San Francisco Chronicle reported that in a testimony to Congress in support of the Raker Bill, engineer Colonel Cosby stated:

> [Hetch Hetchy] is by far the best storage reservoir in that section of the country, and water is so valuable there that they cannot afford to let it run to waste. If you deny the use of it to San Francisco, sooner or later the water will be put to other uses. Somebody will be asking for permission to utilize the Hetch Hetchy valley as a storage reservoir for irrigation purposes.
>
> **(Martin 1914)**

This utilitarian viewpoint of water in service of economic growth was reflected at the highest levels of government. Gifford Pinchot, the then head of the U.S. Forest Service, also testified to Congress in support of the Raker Bill:

> As we all know, there is no use of water higher than the domestic use. Now the fun-damental principle of the whole conservative policy is that of use, to take every part of the land and its resources and put it to that use in which it will best serve the most people.
>
> **(Martin 1914)**

John Muir and other preservationists strongly opposed the Raker Bill. In an impassioned op-ed in the *New York Times* in 1913, Muir stated that building the Hetch Hetchy dam would be "depriving 90,000,000 of people of one of their most priceless possessions [Hetch Hetchy Valley in Yosemite National Park]" (Muir 1913). Figure 5.1 depicts the Tuolumne River flowing through the Hetch Hetchy valley, as shown in a Sierra Club newsletter from 1908 (Taber 1908). Another preservationist, Harriet Monroe, publicly asked, "Will the people of the United States give up for all time one of the wonder valleys of the earth, in order that a single municipality may make a reservoir of it?" (Monroe 1910).

FIGURE 5.1   "Looking Up Hetch Hetchy Valley from Surprise Point," 1908. *Photo credit*:
Isaiah West Taber, Public domain, via Wikimedia Commons.

The answer was "yes." Congress passed the Raker Act in 1913. Water from Hetch
Hetchy reached San Francisco in 1934, after the city had spent over US$100 million
on the project (Hundley 2001).

5.2.2   Water for Ecological Flows

The California State Constitution states that water resources should be "put to beneficial use to the fullest extent of which they are capable," and stipulates that this use
should be "in the interest of the people and for the public welfare" (*California State
Constitution* 1976). Historically, this meant water resources in California were prioritized for commercial and agricultural uses and for urban populations. But in the 1980s,
a landmark legal case caused a shift to the notion that urban Californians are entitled to
all the Sierra snowmelt. Citing the "public trust" doctrine that protects wetlands, shorelines and navigable waters, environmental advocates successfully sued the Los Angeles
Department of Water and Power for causing ecological decline of Mono Lake due to
water diversions to satisfy the needs of Los Angeles's growing population Supreme
Court of California (1983). National Audubon Society v Superior Court; Hager 1982).
This important case overturned Los Angeles' historic water rights to Mono Lake, citing the lake as a "scenic and ecological treasure of national significance" (Hager 1983).

Environmental advocates offered a new paradigm: that more of the water in
California's lakes and rivers should remain there, instead of being appropriated
solely for human use. This ethos emphasized that in addition to allowing urban areas
to thrive, California's water systems and policies must also enable ecological systems and wildlife to survive. Scientists supported this idea, too. In the 1990s and
early 2000s, science emerged that linked the decline of endangered species in the
Sacramento–San Joaquin Delta to diversions of Sierra snowmelt for urban and agricultural uses (Bennett 2005; Moyle et al. 1992; Grimaldo et al. 2009).

In 2009, the California State Legislature passed the Sacramento-San Joaquin Delta Reform Act, which formalized the "co-equal goals" for the state's water systems: providing a more reliable water supply for California as well as protecting, restoring and enhancing the Delta ecosystem (Delta Stewardship Council 2021). As the California State Water Board (Water Board) set regulations about minimum flow levels in the Delta in 2010, it referenced the Mono Lake court decision. The Water Board held that the new flow criteria, which greatly reduced how much water cities and agricultural operations could divert from the Delta, were designed to "protect public trust resources in the Delta ecosystem" (California State Water Resources Control Board and California Environmental Protection Agency 2010; Boxall 2010).

Nearly a century after the Raker Act, there is an ongoing movement to figure out if and how San Francisco can take down the Hetch Hetchy dam, a position that was endorsed by the *New York Times* in 2002 (*New York Times* 2002). Activists illustrated the point by drawing a massive crack on the dam, along with the words, "Free the rivers!" (*Los Angeles Times* 1987) (Figure 5.2).

There is growing awareness that San Francisco and the greater Bay Area can use other local water supplies, including water recycling, groundwater recharge and rainwater capture, to reduce the need for imported water from the Tuolumne River

FIGURE 5.2 The dam at Hetch Hetchy, 2001. In the middle of the dam, you can see the painted zig-zag "crack" and words "Free the rivers" covered with gray paint. *Photo credit:* Richard Luthy.

(*New York Times* 2012). As sympathetic as San Franciscans may be to the values that motivate ecological restoration at Hetch Hetchy valley, the city's residents roundly voted against a 2012 ballot measure even to study the possibility of removing the Hetch Hetchy dam, illustrating residents' uncertainty and unease with jeopardizing their major water supply (Finley 2015).

5.2.3 CLIMATE CHANGE EXACERBATES DROUGHT

The tension about importing water from the mountains to cities in California is set to heighten. As the climate changes and average temperatures rise, Sierra snowmelt and associated river flows are expected to decline. In a high-emissions scenario, climate scientists predict that by the end of the 21st century, Sierra snowpack will average only 20% of what it averaged at the end of the 20th century (Cloern et al. 2011; Ackerly et al. 2018).

The San Francisco Bay Area weathered a severe drought from water-year 2012 to 2016 (Lund et al. 2018). By 2015, Governor Brown passed an executive order mandating California cities reduce their potable water use by 25% compared to what they had used in 2013 (Governor Edmund G. Brown Jr. 2015), which amounted to using 1.2 million acre-feet less water in California cities (California State Water Resources Control Board 2015b). Most of this was achieved due to reductions in outdoor irrigation. In October 2015, the hottest October on record in California to date, the average per-capita water use in the state dropped to 87 gallons per day (California State Water Resources Control Board 2015b).

Bay Area residents can recall their water conservation practices during this time: landscaping and vegetable gardens were replaced with succulents and rocks. Low-flow showerheads and other appliances were made freely available and were widely installed (San Francisco Public Utilities Commission 2014). Parks with grassy fields and landscaped medians were allowed to dry up. People reduced toilet flushing and shortened showers, and encouraged friends and family to follow suit (Richtel 2015).

Even as he declared the drought's end in March 2017, Governor Brown stated, "This drought emergency is over, but the next drought could be around the corner. Conservation must remain a way of life" (Brown 2017). In fact, making conservation a way of life is the first of ten steps in the California Water Action Plan 2016 Update (California Natural Resources Agency, California Department of Food and Agriculture, and California Environmental Protection Agency 2016). Climate projections for the region reinforce the notion that droughts will become more common in the Bay Area (Ackerly et al. 2018).

5.2.4 ACHIEVING ECOLOGICAL GOALS

During the 2012–2016 drought, it became painfully clear that diverting less snowmelt from California's ecosystems would be critical for preserving sensitive habitat. The California State Water Quality Control Board's 2018 Bay Delta Plan noted an 85% net loss in returning adult fall-run Chinook salmon from 1985 to 2017 in the San Francisco Bay Estuary, and cited diversions of the Tuolumne River as part of the problem. Low flows weren't the only problem – it is widely acknowledged that predation by the striped bass, a sport fish introduced to California at the end of the 19th century, is also a threat to salmon in the Delta (Lindley and Mohr 2003) – but water diversions were one of the most concerning. The Bay Delta Plan stated: "While multiple factors are responsible for the [salmon] decline, the magnitude of diversions out of the Sacramento, San Joaquin, and other rivers feeding into the Bay Delta is a major factor in the ecosystem decline." A supporting document to the Bay Delta Plan clearly states as motivation: "The Bay Delta is in ecological crisis" (State Water Resources Control Board and California Environmental Protection Agency 2018).

To protect the ecosystem, the 2018 Bay Delta Plan set standards for the amount of freshwater that should flow in the rivers, including the Tuolumne, that feed the Sacramento-San Joaquin Delta. It states that 40% of the river should be allowed to flow "unimpaired" from February through June through the Delta to the ocean (State Water Resources Control Board and California Environmental Protection Agency 2018). For comparison, only an average of 21% of the Tuolumne was allowed to remain as unimpaired flow in the 25 years between 1984 and 2009 (State Water Resources Control Board and California Environmental Protection Agency 2018). This suggests that the Bay Area would need to cut its diversion of water from the Tuolumne River nearly in half to ensure the survival of the sensitive species and ecosystems which depend upon the river.

While the 2018 Bay Delta Plan took human livelihoods and well-being into account in terms of water allocations, it centered the water requirements of a healthy ecosystem. The Bay Delta Plan acknowledges California's "Human Right to Water" law, stating "It is the policy of the State of California that every human being has the right to safe, clean, affordable, and accessible water adequate for human consumption, cooking, and sanitary purposes" (California State Legislature 2013; State Water Resources Control Board 2016). The Water Board considered this policy, and the 2018 Delta Plan Amendments include a statement that the Water Board "will take actions as necessary to ensure that the implementation of the flow objectives does not impact supplies of water for minimum health and safety needs, particularly during drought periods" (State Water Resources Control Board 2018). Simply stated, the Water Board adopted as policy to restore water flows in the Lower San Joaquin River and its tributaries, including the Tuolumne River, but that is contentious and is not yet implemented (San Francisco Public Utilities Commission 2021b).

Many water managers in the Bay Area are alarmed by the prospects of stark reductions in imported water, particularly during drought years. Nicole Sandkulla, Chief Executive Officer of the Bay Area Water Supply and Conservation Agency (BAWSCA), which purchases imported water from San Francisco from the Tuolumne River, stated that her agency had calculated that

under the State Board's plan, the 1.9 million water users that BAWSCA represents could be forced to reduce average-per-person-per-day water use to 41 gallons a day during a drought, from the recent pre-drought level of 79 gallons per day, and for some people, to 25 gallons per day or less.

(Sandkulla 2020)

As a point of comparison, the World Health Organization (WHO) recommends that the bare minimum amount of clean water necessary per person for basic sanitation is 13–26 gallons per day (United Nations General Assembly 2014). Note that the WHO's guidelines do not include any water for outdoor irrigation of gardens and landscaping – something that typically makes up over half of urban water use in California (Public Policy Institute of California 2016).

5.3   WHAT'S NEXT?

The Bay Area prides itself on its environmental ethos. Most residents would likely agree that it is imperative to leave adequate water in California's rivers to support healthy ecosystems.

So, what would a 50% reduction in urban water use look like? On top of the measures implemented in the last drought, regulations for steep reductions in water use might be difficult to meet by conservation alone. This is especially true in a hotter future, when ample shade trees and urban green space will be critical for keeping cities' temperatures manageable in the dry season.

Efforts to continue improving water conservation and efficient use will continue, and this will likely be a key facet of strategic planning for climate-resilient urban water systems in the Bay Area and across California (Luthy et al. 2020). This may include more expensive options, like repairing leaky pipes – in the Bay Area, between 3% and 15% of the freshwater put into the distribution system is lost to leaks (Krieger 2014). Some amount of drinking water leaking from pipes is unavoidable, because drinking water distribution systems must maintain sufficient pressure for fire-fighting and to keep environmental contaminants (like sewage from other buried pipes) from entering the drinking water pipes (Teunis et al. 2010). Because of this, water-use efficiency efforts may also require more sophisticated methods for remotely detecting leaky pipes and adjusting the pressure accordingly (Xu et al. 2014). There is also a role for improved strategies to support individual behavior change and water conservation attitudes (Willis et al. 2011), through helping motivate individuals to conserve water through in-home displays (Davies et al. 2014) or comparing individuals' water use with that of their peers (Lede et al. 2019).

New water supply options are also on the table. These options include finding ways to capture and store urban rainfall for use in the dry season, and reusing wastewater. Both of these new supplies come with opportunities and challenges, as discussed below. Desalination may also be on the horizon, if technological innovations are developed to make this option cheaper, less energy-intensive and ecologically safer. The future of desalination in the Bay Area will likely entail brackish water rather than ocean water. This is being done already with a 10 MGD brackish groundwater desalination facility by the Alameda County Water District in Newark, and construction has begun on a regional desalination facility that would draw intake water from the Bay in Contra Costa County and produce 6 MGD of fresh water (Bay Area Regional Reliability Project 2017; Bonilla 2021).

Many environmental engineers and water resources managers advocate for "diversifying the water supply portfolio" to make urban water systems more resilient to climate change (Luthy et al. 2020). That entails expanding the menu of water supply options beyond just one source; in the Bay Area, diversifying the water supply portfolio likely will entail continuing to rely on water imports to some extent, while also enhancing conservation efforts, pursuing stormwater capture and investing in water reuse.

5.4 STORMWATER CAPTURE TO AUGMENT URBAN WATER SUPPLIES

Stormwater capture is nothing new in California – along with harvesting snowmelt, many of the state's reservoirs capture rainfall runoff from the surrounding region and store it for future use. Yet expanding reservoirs is not always feasible, and in many parts of California it is highly controversial. Many of the best dam sites have been taken, and there are numerous environmental concerns about constructing more surface water storage (Minton 2008). In addition, climate change is making the state's existing dams more likely to fail because of the increased likelihood of extreme flood events (Mallakpour et al. 2019), and also making it more difficult for them to fulfill the dual role of flood protection and water supply.

A newer practice in California is capturing the stormwater that flows in urban areas. There is a substantial opportunity here: over half of the urban Bay Area region is covered with impervious surfaces that prevent rainfall from soaking into the ground. Instead of allowing rainfall to percolate into the soil, city streets were historically designed to prevent flooding by efficiently channeling all this water into drains so that it could be safely released away from where people live. With a Mediterranean climate where most of the year's rain falls in just a few months of the year, stormwater flows in the Bay Area can be substantial. Even if only a fraction of this stormwater were captured, it could represent a significant contribution to the water supply.

Stormwater capture projects tend to appeal to a range of people in the Bay Area because they can be designed to meet multiple goals: for example, reducing dependence on imported water, controlling floods, preventing surface water pollution, increasing urban green space and providing more area for recreation and wildlife habitat.

Without any additional water treatment, stormwater can be used for non-agricultural landscape irrigation. Yet the practicality of this kind of stormwater capture and direct use is complicated by the fact that the rainfall in the Bay Area is concentrated in the winter when little water is needed for irrigation. If urban stormwater is purified to remove contaminants, it could potentially contribute to potable water supplies or be used for agricultural irrigation.

To make stormwater capture viable as a year-round water supply in the Bay Area, a system for storing the water for later use is critical. One option for stormwater storage is infiltrating stormwater into local aquifers. This is challenging in the Bay Area because the region's steep slopes and clay soils are not conducive to infiltration (Geosyntec Consultants 2011). It is imperative to prevent contaminating the groundwater with polluted stormwater, so pretreatment before infiltration would be necessary. In addition, many local water suppliers do not currently have infrastructure in place to rapidly infiltrate urban runoff. One exception is the Zone 7 Water Agency in the Livermore-Amador Valley that has suitable geology and that is configuring a series of former quarry lakes for purposes of flood control and water banking that includes local stormwater capture and recharge (Zone 7 Water Agency 2014, 2006).

Other Bay Area water managers are also intrigued by stormwater capture because it can be cost-competitive with other new water supply options (San Francisco Public Utilities Commission 2021a). Furthermore, it can reduce adverse effects of large pulses of stormwater on local waterways. Engineered soil geomedia, like specially coated sands, clays or biochar, can be used to purify stormwater during the infiltration process. These geomedia can remove trace organic compounds (e.g., pesticides), nutrients, pathogens and metals (Grebel et al. 2016; Grebel et al. 2013; Mohanty et al. 2018; Ray et al. 2019; Ashoori et al. 2019) to prevent pollution of groundwater. If stormwater storage and infiltration is achieved by means of green infrastructure, stormwater capture can provide additional benefits like urban greening, wildlife habitat and flood prevention. Many local projects to add green infrastructure for stormwater capture in the Bay Area are underway, and more are in development (Figure 5.3). However, very few of the existing green infrastructure features in the Bay Area are designed to augment water supplies with the captured stormwater.

FIGURE 5.3 Construction of a green infrastructure feature for stormwater infiltration and traffic calming near a school zone in Redwood City, California. *Photo credit*: Richard Luthy.

The ways in which stormwater capture is implemented have distinct implications for communities, ecosystems and urban planning. Several different opportunities exist for stormwater capture in the Bay Area. Some of these options center the water supply potential of stormwater: first, stormwater could be captured at a large scale (i.e., from the outfall of a pipe that drains a city's stormwater to the Bay), then diverted to a place where it could be infiltrated to groundwater or stored in reservoirs or cisterns. Similarly, stormwater could be diverted to wastewater treatment plants where water reuse is already being practiced, if capacity exists. Other options for stormwater capture center the multiple benefits possible in these projects, and balance water supply augmentation with other goals like provision of urban green space, community engagement and development of wildlife habitat. These multi-benefit options include integrating stormwater capture and infiltration into green infrastructure projects at parks and along city streets, and household-scale rainwater cisterns.

While each of these types of stormwater capture may play a role in the region's future water supply portfolio, regional coordination is essential to ensure that stormwater capture projects are equitably distributed, effectively placed for maximum benefit and cost-effective. Given the uncertainty associated with future conditions in the Bay Area, maximizing multiple benefits of any long-lived engineered water infrastructure project is prudent. In addition to hedging against the risks of an uncertain future, designing for these multiple benefits can help garner support from a range of key stakeholders and draw funding from a variety of sources.

One example of multi-benefit stormwater capture infrastructure is at Orange Memorial Park in San Mateo County, where three cities and the unincorporated area of San Mateo County are working together to build an innovative stormwater capture system and park. Using funding from the California Department of Transportation, the project will capture and treat water from 6,300 acres in the cities of Colma, Daly City and South San Francisco and store it in cisterns under a new playing field. Some of the water will be used for irrigation, and the rest will be treated and infiltrated into the groundwater basin (Paradigm Environmental 2021; City of South San Francisco 2021). By capturing and treating the stormwater, which is contaminated by urban pollutants like mercury and polychlorinated biphenyls (PCBs), this project will help protect sensitive aquatic habitat in the San Francisco Bay, prevent flooding on local roads and also provide more water for local landscape irrigation.

Another example is the possibility of linking stormwater capture with treatment at water recycling facilities, then pumping the water to recharge basins where it can percolate into the local groundwater supply. In Santa Clara County, a scoping project for the water agency, Valley Water, revealed that it would be possible to capture about 10,000 acre-feet/year of stormwater runoff, send it to an existing water recycling facility for treatment, then pump it approximately 15 miles up to percolation ponds, all at a cost (US$500–700/acre-foot) that is very competitive with existing water supplies (Bradshaw and Luthy 2019). In this case, taking advantage of existing water supply and treatment infrastructure enables cost-effective stormwater capture.

Progress in several key areas could help advance stormwater capture in the Bay Area. These include increased research to fill region-specific data gaps; for example, a clearer understanding of how stormwater capture would affect the ecology of local urban creeks is needed. In Los Angeles, large-scale stormwater capture may result in dramatically reduced streamflow and impair urban riparian ecosystem (Porse and Pincetl 2019) – is the same true in the Bay Area, which averages more annual rainfall than its Southern California neighbor? In addition, the development of metrics for stormwater capture co-benefits to inform decision-making is vital (Luthy et al. 2019). For example, how much flood protection, habitat provision and urban green space would occur from stormwater capture projects that are designed to meet multiple goals in the Bay Area? What other co-benefits of stormwater capture efforts could be monitored and quantified (i.e., improvement in property values, recreational opportunities, pedestrian/bicycle safety, greenhouse gas sequestration, green jobs)?

Any technical solution for stormwater capture would need to be paired with continued integration of stormwater capture into existing local and regional planning documents, along with analysis of funding options for stormwater capture in the Bay Area.

One of the biggest challenges to advancing stormwater capture for urban water supply in the Bay Area is the current institutional arrangement of separate entities managing flood protection, pollution control and water supply. Building bridges between these separate agencies so they can work together to plan and implement stormwater capture, along with members of local communities who can prioritize the benefits associated with the project, is a critical piece of the puzzle (Harris-Lovett et al. 2019).

5.5 WATER REUSE

As freshwater becomes considered more of a scarce commodity in California, water reuse – the practice of recycling water in sewage for productive purposes – becomes more attractive (Luthy et al. 2020). There is broad interest in water reuse across the state and the nation, especially because of its potential role in diversifying urban water portfolios to make them more resilient to a changing climate (U.S. EPA. 2019; Hering et al. 2013). Advocates for water reuse see wastewater as a resource that should be put to productive use, particularly because it can stretch the utility of imported water by using it more than once. As of 2015, about 93% of wastewater from Bay Area residents was discharged into the San Francisco Bay and the Pacific Ocean, with only 7% reused (California State Water Resources Control Board 2015a).

In the Bay Area, water managers and other stakeholders are interested in increasing the amount of water reused because it can allow for what is perceived as appropriate use of critical water resources, provide environmental benefit by potentially reducing the region's demand for imported water and burnish a "green" reputation for the implementing organization. Furthermore, water reuse is attractive to some water managers because it can efficiently meet regulations on wastewater discharge, for example, by diverting wastewater for irrigation that would otherwise require more stringent (and costly) treatment before it is released to ecologically sensitive portions of the San Francisco Bay (Harris-Lovett et al. 2019).

5.5.1 Non-potable Reuse

Most urban water reuse in California is currently employed for non-potable purposes, like for landscape irrigation, industrial cooling or habitat restoration. Currently, a whopping 10% of all water reuse in the Bay Area goes to irrigating golf courses (California State Water Resources Control Board 2015a). In the Bay Area, non-potable water reuse, where the nutrients in the wastewater could be beneficially used for irrigation, could potentially save the region on the order of a billion dollars because it could prevent the need for costly upgrades to wastewater treatment plants to remove nutrients from wastewater (Harris-Lovett et al. 2018). Regulations and technologies for non-potable municipal water reuse are well-established in California (California State Water Resources Control Board 2016).

Yet in the Bay Area, many of the "low hanging fruits" of non-potable water reuse (the large water users near sea level) have already been taken. In addition, the cost of constructing pipes to transport non-potable water can be prohibitively expensive, and in some places logistically impossible (Bischel et al. 2012). Another challenge for expanding municipal-scale non-potable water reuse in the Bay Area will be doing so in ways that are aligned with regional values – for example, some have expressed a desire not to use recycled water to support more golf courses or industrial cooling for the region's oil refineries (Harris-Lovett et al. 2019). New uses for non-potable water reuse specific to climate adaptation, like irrigation of urban forests or vegetated levees for flood protection, may be more well-supported by members of the public.

Non-potable water reuse can also happen at a building scale, for example, by reusing household graywater or foundation drainage water for toilet flushing. Adding non-potable water reuse systems in large new buildings, particularly in developments like office parks or tech-industry campuses, is likely the most feasible option, since retrofitting old buildings can be logistically infeasible. Businesses may be valuable allies in creating decentralized infrastructure for water reuse, as they can develop and implement it way faster than municipal governments can, and such developments can burnish a company's "green" image (Harris-Lovett et al. 2019). The creation of uniform standards for onsite water reuse systems would help improve the ability of these systems to operate safely and reliably over time. One of the challenges with onsite water reuse is ensuring it is designed, implemented and maintained in a manner that promotes social equity. This is a topic that has been explored much more deeply about decentralized (renewable) energy systems (see, e.g., Alstone et al. 2015; Adil and Ko 2016), but is still understudied in the area of onsite water infrastructure.

As a climate adaptation strategy, building- or neighborhood-scale water reuse may hold promise in the Bay Area, because so many of the region's existing wastewater treatment plants are located at sea level and at dire risk of flooding in the future (Heberger et al. 2009). Reusing water from higher elevations in the sewershed, further away from the wastewater treatment plants, provides several different benefits. First, it allows for less pumping costs associated with taking the treated water to the place where it will be used (Gikas and Tchobanoglous 2009), and lower costs associated with building pipelines to bring recycled water from the place where it is produced to where it will be used (Bischel et al. 2012). Second, onsite water reuse ensures water infrastructure investments are further from the shoreline and hence more resilient to a future with higher sea levels. Finally, sewers at sea level are subject to infiltration of salty water into the pipes, leading to the need for more expensive, energy-intensive treatment for water reuse, and this problem will be exacerbated by rising sea levels. Onsite, up-slope reuse avoids the problem of brackish water infiltration into the sewers, leading to potentially less expensive treatment.

Several large technology companies in the San Francisco Bay Area as well as the City of San Francisco are pursuing onsite water reuse in some of their large buildings. One example of this is Facebook, which created an onsite water reuse system for flushing toilets and irrigating their grounds (Figure 5.4). One of the main motivations for constructing such a system was to burnish the company's "green" reputation (Harris-Lovett et al. 2019).

As onsite water infrastructure implemented by private companies or other entities becomes more common, collaboration between these businesses and municipal agencies is critical to harmonize planning between the two for cost-efficiency and to avoid stranded investments. Urban water reuse systems that contain both centralized and decentralized elements must be carefully coordinated to prevent unwanted technical impacts of decentralized projects on the municipal infrastructure, for example, by depriving it of flows, or increasing salinity (Sapkota et al. 2015). In addition, decentralized water reuse requires new institutional arrangements to promote fair governance and cost-sharing relates to both the decentralized projects and the municipal infrastructure (Hoffmann et al. 2020).

FIGURE 5.4 Onsite water reuse equipment at the Facebook corporate campus (left) produces freshwater to irrigate rooftop gardens where staff members can eat lunch and relax (right). *Photo credits*: Richard Luthy.

5.5.2 POTABLE REUSE

Water engineers in California have also expressed increasing interest in potable water reuse, which entails purifying sewage to drinking water standards and introducing it into a drinking water aquifer, reservoir or into the drinking water distribution system (Binz et al. 2016). While some parts of Southern California have employed potable water reuse for nearly 50 years (Harris-Lovett and Sedlak 2015), it is new to the Bay Area. Public reception to the notion of potable water reuse projects varies significantly across California communities (Harris-Lovett et al. 2015), and some communities may be more suitable for it than others based on the potential for controlling entry of harmful chemicals into the sewer system (Harris-Lovett and Sedlak 2020). Opinions about potable water reuse also vary widely within the metropolitan San Francisco Bay Area. In the Tri-Valley area of Contra Costa County, the Pleasanton city council recently voted against even studying the feasibility of potable water reuse for their city, with the mayor Karla Brown stating, "If you have a brand new baby, do you want to feed that baby potable reuse water? I don't (Baum 2021)." On the other side of the spectrum, the Silicon Valley Advanced Water Purification System in the South Bay currently purifies wastewater for irrigation and other non-potable uses, but the facility is a pilot project to demonstrate the feasibility of potable water reuse and the water agency is planning to expand the facility to serve that purpose in the future (Figure 5.5) (Valley Water 2019). The Silicon Valley Advanced Water Purification System is also designed such that it can easily be re-purposed

FIGURE 5.5 Banks of reverse osmosis water treatment equipment comprise one of the steps in producing high-quality water from wastewater at the Silicon Valley Advanced Water Purification System in San Jose, California. *Photo credit:* Richard Luthy.

to produce potable water in the event of a drought (Bay Area Regional Reliability Project 2017).

Standardized public health regulations for potable water reuse in California are more recent, with some still in development (California State Water Resources Control Board 2021). In 2018, California passed regulations specifying standards for adding recycled water to drinking water reservoirs. These standards include the percent of recycled water (by volume) that can be added to a reservoir, as well as the concentrations of pathogens and chemical contaminants it can contain (California State Water Resources Control Board 2018). These regulations for potable water reuse are a result of a great deal of institutional work, including intensive strategizing and lobbying, by the state's largest water agencies and a water reuse industry advocacy organization (Binz et al. 2016). Having clear regulations for potable water reuse builds cultural legitimacy for the practice as well as streamlines the permitting process for reuse projects (Harris-Lovett et al. 2015; Mukherjee and Jensen 2020).

5.5.3 COLLABORATION TO ADVANCE WATER REUSE

To date, most water reuse projects in the Bay Area are conceived and managed on a case-by-case basis. Yet a regional approach to water reuse in the Bay Area would have many advantages: for example, by ensuring the necessary coordination between water supply and wastewater agencies for planning and implementing water reuse projects. Regional collaboration could also allow water managers to most cost-effectively take advantage of site-specific opportunities for water reuse (Harris-Lovett et al. 2019).

If there is one thing we have learned from previous droughts in California, it's this: people think a lot more about the importance of water, and are more willing to invest in it, when it's not raining. Planning for water reuse in advance of droughts can enable water reuse projects to quickly come to fruition when opportunities for funding are available.

Even as water reuse provides a means of diversifying the Bay Area's water portfolio so that the region could be less reliant on imported water, there is tension about whether that will be the reality of water reuse. Legislation in the Bay Area requires that new housing developments prove they have access to adequate drinking water before construction (Hanak 2010) – and there is concern that in reality, water reuse may actually allow increased development rather than serving to replace imported flows (Harris-Lovett et al. 2019). Regulatory agencies may be able to have a role in ensuring that water reuse projects serve to make the region more resilient to climate change by reducing imported water.

Advancing water reuse in the Bay Area depends upon addressing specific gaps and barriers in the region. Water and wastewater agencies, regulatory agencies, environmental advocates and community groups can collaborate to develop a "hierarchy of uses" for water reuse, which can be used to incentivize those projects that are aligned with regional values. They can also advocate for developing a "sewershed management" approach to ensure that industrial pollutants and other chemicals that are difficult to remove in water treatment processes do not contribute to wastewater destined for potable reuse (Harris-Lovett and Sedlak 2020). Regulators can consider laws to ensure that water reuse projects result in environmental benefit (for example, by reducing water imports to the region). They can also modernize non-potable water reuse requirements to remove unintended disincentives to reuse.

Increasing water reuse in the Bay Area will require that water utilities in the region continue moving from a mindset of selling water to one of integrated water management, and will necessitate building bridges between traditionally separate agencies and professionals in diverse roles. Integrating resilience to a changing climate, social equity and environmental benefit into the metrics of success for water infrastructure will help determine the appropriate water reuse projects for the region.

5.6 CHANGING REGULATIONS, CHANGING PARADIGMS

Water reuse and stormwater capture, as described above, may be able to work synergistically to augment water supplies in California. This could be done by directing urban stormwater to wastewater treatment plants, to enable it to be treated and reused. This would benefit water supply by boosting the volume of water available for reuse while also preventing polluted stormwater from running unimpeded into delicate surface waters, i.e., the San Francisco Bay Estuary (Harris-Lovett et al. 2019). Diverting stormwater to wastewater treatment plants for purification and reuse could be especially helpful for mitigating the effects of contaminants in the so-called "first flush", which is the stormwater that flows into an aquatic ecosystem after the first rains of the year in a Mediterranean climate wash the accumulated pollutants off the streets (Lee et al. 2004). Pilot studies in southern part of the San Francisco Bay Area have evaluated the potential for diverting stormwater to wastewater treatment plants to prevent pollution (Santa Clara Valley Urban Runoff Pollution Prevention Program 2015), but none to our knowledge have yet evaluated the potential for water supply augmentation in the region.

Long ago, urban water systems in California originally combined their sewer networks with storm drains to prevent flooding and remove wastewater. In the Bay Area, most of these combined storm drain/sewer systems were decoupled in order to prevent large wintertime rains from exceeding wastewater treatment plant capacity, resulting in discharges of untreated sewage (known as combined sewer overflows, or CSOs). Of the more than 40 wastewater districts in the Bay Area, only San Francisco currently has a combined sewer system. Recombining sewer and stormwater drainage systems in the interest of boosting water supplies available for treatment and reuse would have to entail smart flow control devices that would divert stormwater away from the sewer (and back to being directly discharged to the Bay) if flows were expected to exceed wastewater treatment capacity.

An example of this is occurring in Southern California, where the City of Los Angeles is beginning construction on infrastructure intended to direct dry-season stormwater flows, the so-called "urban drool" that is the polluted runoff from irrigation, car washing and other summertime activities, to the local wastewater treatment plant (City of Los Angeles Bureau of Engineering 2021). This will also help prevent litter and other urban pollutants from harming water quality at the region's beaches. Similar projects in the Bay Area would be beneficial.

5.6.1 TAKING WATER REUSE TO THE NEXT LEVEL

As notions about resilient water systems affect design and planning, these ideas have also been reflected in regulatory processes. Perhaps one of the best illustrations of the changing paradigm for urban water systems in California is the Herzberg bill, SB 332, which was proposed to the California State legislature in 2019. This bill declared that releasing treated wastewater effluent to the ocean was a "waste and unreasonable use of water," stating that nearly all ocean discharges of wastewater in California should be halted by 2040. Instead, 95% of the state's treated wastewater effluent, estimated to be more than a trillion gallons per day (California SB332 2019), should be productively reused (Hertzberg and Weiner 2019). The bill was sent back to the state Senate for revisions in February 2020, and its fate is still uncertain. Despite that, it has raised a stir among the major stakeholders in communities of wastewater managers, water suppliers and water reuse advocates. Many of them oppose the bill because these rigid limits, to be adopted on a relatively short timeline, would make it difficult and costly to implement across the range of wastewater treatment plants in the state (Gauger et al. 2019; West 2019). Regardless, the Hertzberg bill is a signpost for the future of wastewater treatment in California – eventually, it is likely to become more of a resource recovery operation than a system for treating wastewater for safe disposal (Hering et al. 2013; Daigger 2009).

5.6.2 COPING WITH UNCERTAINTY

Just as future regulations about wastewater disposal remain uncertain, future regulations associated with imported water supply options also remain unclear. Whether or not the California State Water Quality Control Board succeeds in ensuring 40% of the Tuolumne River remains as "in-stream" flows – they have stated they intend to do so by 2022, but numerous lawsuits have been brought to them to try to revoke the changing requirements (San Francisco Public Utilities Commission 2021b) – will have large ramifications for Bay Area water suppliers who currently rely heavily upon that imported water supply.

The first tool for coping with this uncertainty is planning. At a state-wide level, new guidance from the California Department of Water Resources on urban water supply management mandates urban water agencies to model their plans for a minimum of five consecutive dry years, rather than the three years as required previously (California Department of Water Resources 2021a). Locally, Bay Area water agencies are parsing out the details of what this might mean for their unique context. For example, in addition to accounting for uncertainties about the amount and timing of natural flows in the Tuolumne River due to the vagaries of weather and a changing climate, the San Francisco Public Utilities Commission (SFPUC) prepared separate plans for whether or not the amount of water they are allowed to divert from the river changes. There is so much uncertainty about the Tuolumne that SFPUC presents two scenarios – one with no restrictions on unimpaired flow, and one with 40%

unimpaired flow (San Francisco Public Utilities Commission 2021b). Along with consideration of engineered water supply alternatives like water reuse and desalination of brackish water in the San Francisco Bay Estuary to meet the demands in these scenarios, the San Francisco Public Utilities Commission is also considering new approaches for sharing water.

This leads to the second tool for coping with uncertainty: partnerships. Eight of the largest Bay Area water agencies, including SFPUC, are planning to build pipes between water agencies to share water resources regionally in the event of a severe drought (Bay Area Regional Reliability Project 2017). Some of these resources will likely entail new supplies of recycled water for potable reuse in Silicon Valley – which will also require partnerships between the wastewater producers and the water supply agencies (Bay Area Regional Reliability Project 2017). SFPUC is also hoping to develop "groundwater banking" agreements with other major water users with rights to the Tuolumne River, namely the Turlock and Modesto Irrigation Districts, which actually have senior water rights to the Tuolumne as established by the 1913 Raker Act (San Francisco Public Utilities Commission 2021b). A water banking agreement with the Turlock and Modesto Irrigation Districts would allow the irrigators to receive additional shares of river water in wet years, which would offset the irrigators' use of groundwater. The irrigators could then use the "banked" groundwater that remained underground in a subsequent dry year, thereby allowing San Francisco to use more imported water in that year (San Francisco Public Utilities Commission 2021b).

To expand the groundwater banking opportunity for the Bay Area, SFPUC or other Bay Area water agencies may be able to help finance the identification of areas in Turlock or Modesto where the geology is especially well-suited for additional infiltration, as well as the infrastructure to infiltrate additional high-flow water from the Tuolumne River there. This would be especially helpful, since in about half of the wet and above-normal rainfall years, the Tuolumne River flows much higher than is needed for combined urban, agricultural and ecological use (Modesto Irrigation District 2020). Such a plan is under discussion in a proposed "Voluntary Agreement" about the Tuolumne River, supported by Modesto and Turlock Irrigation Districts and SFPUC, which they intend to use to supplant the Water Board's ruling about leaving in-stream flows in the Tuolumne (Carlson 2019).

Finally, developing multi-benefit water infrastructure can be an effective way to hedge against the uncertainty associated with climate change impacts (Harris-Lovett et al. 2019; Rohde, Reynolds and Howard 2020). Multi-benefit water infrastructure can end up being more cost-efficient overall, while also broadening support for a new project and helping build a coalition of stakeholders to champion it (Diringer et al. 2019).

5.7   CONCLUSIONS: A REGIONAL APPROACH FOR EQUITABLE DESIGN AND IMPLEMENTATION

Many changes must be made to California's urban water systems in order for them to be more resilient to a changing climate. Less dependence on imported water from the Sierra Nevada will make cities like the metropolitan Bay Area more able to withstand drought, while also protecting the wildlife species that depend on snow-melt-fed rivers for their survival. Stormwater capture and water reuse may be part of the solution.

These solutions are best achieved through regional collaboration. Rather than having each of the more than 50 water supply agencies in the San Francisco Bay Area (Sommer 2014) try to develop their own separate supplies, working together can make a huge difference in terms of taking advantage of regional pilot project opportunities, taking advantage of economies of scale for cost-effective investments and providing access to new sources of funding. Regional collaboration can also forward the goal of equitable water service provision across the Bay Area by ensuring that the smaller water agencies with less resources are included in the plans and projects.

Developing stormwater capture and water reuse as new water supplies entails effective networks among agencies that traditionally work in very separate silos, such as drinking water, stormwater and wastewater management. Some of this kind of regional collaboration is underway in the Bay Area. For example, water agencies, wastewater agencies, researchers and nonprofit organizations have created the Bay Area One Water Network, a forum called for sharing ideas and strategizing next steps specifically about options for water supply resilience in the Bay Area ("Bay Area One Water Network" 2021).

Designing and implementing resilient water systems, like many other forms of climate adaptation, risks deepening existing social inequalities if specific action is not taken to deliberately center equity in the process (Rodina et al. 2017). Key questions for consideration include: How will decisions about new water infrastructure developments be made, and who will be included in the planning process? How can the goals of the project be made reflective of diverse stakeholder groups (including low-income communities, people of color or other marginalized groups)? How will decisions about Bay Area water infrastructure affect geographically distant populations, as well as future generations? Who will pay for and who will benefit from new water investments?

In addition, support for water system adaptation at the state and federal levels can help ensure equitable distribution of resources across regions, so that innovative strategies are not only implemented in the wealthiest areas with the most available time and money to devote to it. This has recently begun, for example, through directives from the California Water Resilience Portfolio Initiative and the National Water Reuse Action Plan (California Natural Resources Agency, California Environmental Protection Agency and California Department of Food and Agriculture 2020; U.S. EPA 2019).

None of this is easy. Getting diverse stakeholders from different agencies, organizations and community groups together to strategize, plan and implement new water infrastructure requires time, patience and resources. But given the stark impacts of the climate crisis on California's urban water supplies, it is essential. By centering equity and cooperating regionally to develop water systems that are responsive to local cultures, concerns and conditions, the Bay Area can develop resilient water systems that can withstand a changing climate and future droughts.

LITERATURE CITED

Ackerly, D., Jones, A., Stacey, M., & Riordan, B. (2018). California's fourth climate change assessment: San Francisco Bay area summary report. CCCA4-SUM-2018–005. Retrieved from https://www.climateassessment.ca.gov.

Adil, A. M., & Ko, Y. (2016). Socio-technical evolution of decentralized energy systems: A critical review and implications for urban planning and policy. *Renewable and Sustainable Energy Reviews, 57*(May), 1025–1037. doi: 10.1016/j.rser.2015.12.079.

Alstone, P., Gershenson, D., & Kammen, D. M. (2015). Decentralized energy systems for clean electricity access. *Nature Climate Change, 5*(4), 305–314. doi: 10.1038/nclimate2512.

Arcuni, P. (2020). As April begins, California's snowpack is about half of normal. *KQED.* Retrieved from https://www.kqed.org/science/1960807/as-april-begins-californias-snowpack-is-about-half-of-normal..

Ashoori, N., Teixido, M., Spahr, S., LeFevre, G. H., Sedlak, D. L., & Luthy, R. G. (2019). Evaluation of pilot-scale biochar-amended woodchip bioreactors to remove nitrate, metals, and trace organic contaminants from urban stormwater runoff. *Water Research, 154*, 1–11.

Association of Bay Area Governments. (2021). Our members. Retrieved from https://abag.ca.gov/about-abag/what-we-do/our-members.

Baum, J. (2021). Pleasanton city council drops pursuit for potable water. *Pleasanton Weekly (online edition).* California: Pleasanton. February 2021. Retrieved from https://www.pleasantonweekly.com/news/2021/02/03/pleasanton-city-council-drops-pursuit-for-potable-water.

Bay Area One Water Network. (2021). Bay area one water network. Retrieved from https://www.bayareawater.org.

Bay Area Regional Reliability Project. (2017). Bay area regional reliability drought contingency plan final. Retrieved from https://www.bayareareliability.com/uploads/BARR-DCP-Final-12.19.17-reissued.pdf.

Bennett, W. A. (2005). Critical assessment of the delta smelt population in the San Francisco estuary, California. *San Francisco Estuary and Watershed Science, 3*(2). doi: 10.15447/sfews.2005v3iss2art1.

Binz, C., Harris-Lovett, S., Kiparsky, M., Sedlak, D. L., & Truffer, B. (2016). The thorny road to technology legitimation—Institutional work for potable water reuse in California. *Technological Forecasting and Social Change, 103*, 249–263.

Bischel, H. N., Simon, G. L., Frisby, T. M., & Luthy, R. G. (2012). Management experiences and trends for water reuse implementation in northern California. *Environmental Science and Technology, 46*(1), 180–188. doi: 10.1021/es202725e.

Bonilla, R. (2021). City of Antioch breaks ground on historic, first desalination project in delta. *Contra Costa Herald.* February 20, 2021. Retrieved from https://contracostaherald.com/city-of-antioch-breaks-ground-on-historic-first-desalination-project-in-delta/.

Boxall, B. (2010). Report urges less use of delta water; the study could lay the groundwork for limits on Southern California supply. *Los Angeles Times,* July 22, 2010. Retrieved from http://search.proquest.com/docview/625868568/abstract/1F183945B00840D4PQ/20.

Bradshaw, J. L., & Luthy, R. G. (2019). Researching cost-effective opportunities for beneficial use of stormwater in Santa Clara county. *Valley Water,* September 25.

Brown, E. G. (2017). Governor Brown lifts drought emergency, retains prohibition on wasteful practices. Retrieved from https://www.ca.gov/archive/gov39/2017/04/07/news19748/index.html.

California Department of Water Resources. (2021a). Urban water management plan Guidebook 2020. Retrieved from https://water.ca.gov/-/media/DWR-Website/Web-Pages/Programs/Water-Use-And-Efficiency/Urban-Water-Use-Efficiency/Urban-Water-Management-Plans/Final-2020-UWMP-Guidebook/UWMP-Guidebook-2020---Final-032921.pdf.

California Department of Water Resources. (2021b). Statewide snowpack well below normal as wet season winds down. Retrieved from https://water.ca.gov/News/News-Releases/ 2021/April-21/Statewide-Snowpack-Well-Below-Normal-as-Wet-Season-Winds-Down.

California Natural Resources Agency, California Department of Food and Agriculture, & California Environmental Protection Agency. (2016). California water action plan 2016 update. Retrieved from https://resources.ca.gov/CNRALegacyFiles/docs/california_ water_action_plan/Final_California_Water_Action_Plan.pdf.

California Natural Resources Agency, California Environmental Protection Agency, & California Department of Food and Agriculture. (2020). California water resilience portfolio draft. Retrieved from http://waterresilience.ca.gov/wp-content/uploads/2020/ 01/California-Water-Resilience-Portfolio-2019-Final2.pdf.

California SB332. (2019). Retrieved from https://trackbill.com/bill/california-senate-bill-332-wastewater-treatment-recycled-water/1695900/

California State Constitution. (1976). Article X Section 2. Retrieved from https://leginfo.leg-islature.ca.gov/faces/codes_displayText.xhtml?lawCode=CONS&article=X.

California State Legislature. (2013). California water code. 106.3. Retrieved from https:// leginfo.legislature.ca.gov/faces/codes_displaySection.xhtml?lawCode=WAT&section-Num=106.3.

California State Water Resources Control Board. (2015a). 2015 California municipal waste-water recycling survey. Retrieved from https://www.waterboards.ca.gov/water_issues/ programs/grants_loans/water_recycling/munirec.shtml.

California State Water Resources Control Board. (2015b). California's cumulative water savings continue to meet governor's ongoing conservation mandate. Retrieved from https://www.waterboards.ca.gov/press_room/press_releases/2015/pr120115_oct_con-servation.pdf.

California State Water Resources Control Board. (2016). Water reclamation requirements for recycled water use. Order WQ 2016–0068-DDW. Retrieved from https://www.waterboards .ca.gov/board_decisions/adopted_orders/water_quality/2016/wqo2016_0068_ddw.pdf.

California State Water Resources Control Board. (2018). Regulations related to recycled Water. California code of regulations. Vol. Title 22. Retrieved from https://www. waterboards.ca.gov/drinking_water/certlic/drinkingwater/documents/lawbook/ RWregulations_20181001.pdf.

California State Water Resources Control Board. (2021). Regulating direct potable reuse in California. Retrieved from https://www.waterboards.ca.gov/drinking_water/certlic/ drinkingwater/direct_potable_reuse.html.

California State Water Resources Control Board and California Environmental Protection Agency. (2010). *Development of flow criteria for the Sacramento-San Joaquin Delta ecosystem*. Sacramento, CA.

Carlson, K. (2019). Voluntary agreements shared with state water board. Will they replace disputed flow plan? *Modesto Bee*. March 1, 2019. Retrieved from https://www.modbee. com/news/article227010034.html.

City of Los Angeles Bureau of Engineering. (2021). Los Angeles river and Arroyo Seco low flow diversion. Retrieved from https://eng.lacity.org/lalowflowdiversion. Checked.

City of South San Francisco. (2021). Orange memorial park improvements. Retrieved from https://www.ssf.net/government/construction/orange-memorial-park-improvement.

Cloern, J. E., Knowles, N., Brown, L. R., Cayan, D., Dettinger, M. D., Morgan, T. L., et al. (2011). Projected evolution of California's San Francisco Bay-delta-river system in a century of climate change. *PLOS ONE, 6*(9), e24465. doi: 10.1371/journal.pone.0024465.

Daigger, G. (2009). Evolving urban water and residuals management paradigms: Water reclamation and reuse, decentralization, and resource recovery. *Water Environment Research, 81*(8), 809–823. doi: 10.2175/106143009X425898.

Davies, K., Doolan, C., Van Den Honert, R., & Shi, R. (2014). Water-saving impacts of smart meter technology: An empirical 5 year, whole-of-community study in Sydney, Australia. *Water Resources Research, 50*(9), 7348–7358. doi: 10.1002/2014WR 015812.

Delta Stewardship Council. (2021). Delta stewardship council: Frequently asked questions. Retrieved from https://deltacouncil.ca.gov/frequently-asked-questions.

Diffenbaugh, N. S., Swain, D. L., & Touma, D. (2015). Anthropogenic warming has increased drought risk in California. *Proceedings of the National Academy of Sciences, 112*(13), 3931–3936.

Diringer, S., Thebo, A., Cooley, H., Shimabuku, M., Wilkinson, R., & Bradford, M. (2019). *Moving toward a multi-benefit approach for water management.* Oakland, CA: Pacific Institute. Retrieved from https://pacinst.org/wp-content/uploads/2019/04/moving-toward-multi-benefit-approach.pdf.

Finley, A. (2015). Cross country: Hetch Hetchy makes San Franciscans a touch tetchy. *Wall Street Journal, Eastern Edition*, May 9, 2015. Retrieved from http://search.proquest.com/docview/1679659585/abstract/2507AA29EBD14528PQ/1.

Gauger, J., Blacet, D., Gervase, R., Quinonez, A., & Dolfie, D. (2019). SB 332 (Hertzberg & Wiener): Oppose. Retrieved from https://casaweb.org/wp-content/uploads/2020/10/5-6-19-SB-332-Opposition-Letter-Approps.pdf.

Geosyntec Consultants. (2011). *Harvest and use, infiltration and evapotranspiration feasibility/infeasibility criteria report.* Menlo Park, CA: Bay Area Stormwater Management Agencies Association.

Gikas, P., & Tchobanoglous, G. (2009). The role of satellite and decentralized strategies in water resources management. *Journal of Environmental Management, 90*(1), 144–152.

Governor Edmund G. Brown Jr. (2015). *Executive order, B-29–B-15.* Sacramento, the capital of California.

Grebel, J. E., Charbonnet, J. A., & Sedlak, D. L. (2016). Oxidation of organic contaminants by manganese oxide geomedia for passive urban stormwater treatment systems. *Water Research, 88*, 481–491.

Grebel, J. E., Mohanty, S. K., Torkelson, A. A., Boehm, A. B., Higgins, C. P., Maxwell, R. M., … Sedlak, D. L. (2013). Engineered infiltration systems for urban stormwater reclamation. *Environmental Engineering Science, 30*(8), 437–454.

Grimaldo, L. F., Sommer, T., Van Ark, N., Jones, G., Holland, E., Moyle, P. B., … Smith, P. (2009). Factors affecting fish entrainment into massive water diversions in a tidal freshwater estuary: Can fish losses be managed? *North American Journal of Fisheries Management, 29*(5), 1253–1270. doi: 10.1577/M08-062.1.

Hager, P. (1982). 'Trust' issue raised in mono fight: State water law could change if court accepts argument MONO: New issue raised. *Los Angeles Times (1923–1995)*, May 11, 1982, sec. Part II. Retrieved from http://search.proquest.com/docview/153143108/abstract/B4F9CAE2CEC4228PQ/1.

Hager, P. (1983). Court clears way for ruling on mono Water. *Los Angeles Times (1923–1995)*, February 18, 1983, sec. Part. I.

Hanak, E. (2010). Show me the water plan: Urban water management plans and California's water supply adequacy laws. *Golden Gate University Environmental Law Journal, 4*(1). Retrieved from https://waterinthewest.stanford.edu/sites/default/files/related_documents/Show_Me_Water_Plan.pdf.

Harris-Lovett, S., Baker, K., Luthy, R., & Sedlak, D. L. (2019). *Advancing water reuse in the bay area: Integrating water reuse into a regional approach to water management.* Bay Area One Water Network. San Francisco, CA.

Harris-Lovett, S., Baker, K., Mayo, M., Sedlak, D., & Luthy, R. (2019). *Stormwater capture to augment water supplies in the San Francisco Bay Area: Challenges, opportunities, and next steps.* Bay Area One Water Network. San Francisco, CA.

Harris-Lovett, S., Binz, C., Sedlak, D. L., Kiparsky, M., & Truffer, B. (2015). Beyond user acceptance: A legitimacy framework for potable water reuse in California. *Environmental Science and Technology, 49*(13), 7552–7561. doi: 10.1021/acs.est.5b00504.

Harris-Lovett, S., Lienert, J., & Sedlak, D. (2018). Towards a new paradigm of urban water infrastructure: Identifying goals and strategies to support multi-benefit municipal wastewater treatment. *Water, 10*(9), 1127.

Harris-Lovett, S., Lienert, J., & Sedlak, D. (2019). A mixed-methods approach to strategic planning for multi-benefit regional water infrastructure. *Journal of Environmental Management, 233*(March), 218–237. doi: 10.1016/j.jenvman.2018.11.112.

Harris-Lovett, S., & Sedlak, D. L. (2015). The history of water reuse in California. In A. Lassiter (Ed.), *Sustainable water: Challenges and solutions from California*. Berkeley, CA: University of California Press.

Harris-Lovett, S., & Sedlak, D. (2020). Protecting the Sewershed. *Science, 369*(6510), 1429–1430. doi: 10.1126/science.abd0700.

Heberger, M., Cooley, H., Herrera, P., Gleick, P. H., & Moore, E. (2009). The impacts of sea-level rise on the California Coast. California Climate Change Center CEC-*500-2009-024-F*. Retrieved from http://dev.cakex.org/sites/default/files/CA%20Sea%20Level%20Rise%20Report.pdf.

Hering, J. G., Waite, T. D., Luthy, R. G., Drewes, J. E., & Sedlak, D. L. (2013). A changing framework for urban water systems. *Environmental Science and Technology, 47*(19), 10721–10726. doi: 10.1021/es4007096.

Hertzberg, R., & Weiner, S. (2019). Wastewater treatment: Recycled Water. Retrieved from https://leginfo.legislature.ca.gov/faces/billVersionsCompareClient.xhtml?bill_id=201920200SB332.

Hoffmann, S., Feldmann, U., Bach, P. M., Binz, C., Farrelly, M., & Frantzeskaki, N. (2020). A research agenda for the future of urban water management: Exploring the potential of nongrid, small-grid, and hybrid solutions. *Environmental Science and Technology, 54*(9), 5312–5322.

Hundley, N. (2001). *The great thirst: Californians and water - A history*. University of California Press. Oakland, CA: University California Press.

Krieger, L. (2014). California drought: Bay area loses billions of gallons to leaky pipes. *The Mercury News*. Retrieved from https://www.mercurynews.com/2014/08/16/california-drought-bay-area-loses-billions-of-gallons-to-leaky-pipes/.

Lede, E., Meleady, R., & Seger, C. R. (2019). Optimizing the influence of social norms interventions: Applying social identity insights to motivate residential water conservation. *Journal of Environmental Psychology, 62*, 105–114. doi: 10.1016/j.jenvp.2019.02.011.

Lee, H., Lau, S.-L., Kayhanian, M., & Stenstrom, M. K. (2004). Seasonal first flush phenomenon of urban stormwater discharges. *Water Research, 38*(19), 4153–4163.

Lindley, S. T., & Mohr, M. S. (2003). Modeling the effect of striped bass (Morone saxatilis) on the population viability of Sacramento River winter-run chinook salmon (*Onchorhynchus Tshawytscha*). *Fishery Bulletin, 101*(2), 321–331.

L.A. Times Archives. (1987). 'Really a Work of Art': Vandal paints crack on park dam. Retrieved from https://www.latimes.com/archives/la-xpm-1987-07-21-mn-5359-story.html.

Lund, J., Medellin-Azuara, J., Durand, J., & Stone, K. (2018). Lessons from California's 2012–2016 drought. *Journal of Water Resources Planning and Management, 144*(10), 04018067. doi: 10.1061/(ASCE)WR.1943-5452.0000984.

Luthy, R. G., Sharvelle, S., & Dillon, P. (2019). Urban stormwater to enhance water supply. *Environmental Science and Technology, 53*(10), 5534–5542. doi: 10.1021/acs.est.8b05913.

Luthy, R. G., Wolfand, J. M., & Bradshaw, J. L. (2020). Urban water revolution: Sustainable water futures for California cities. *Journal of Environmental Engineering, 146*(7), 04020065. doi: 10.1061/(ASCE)EE.1943-7870.0001715.

Mallakpour, I., AghaKouchak, A., & Sadegh, M. (2019). Climate-induced changes in the risk of hydrological failure of major dams in California. *Geophysical Research Letters, 46*(4), 2130–2139.

Martin, V. (1914). How the Raker bill affects Hetch-Hetchy, San Francisco and state: Facts regarding mountain supply history of city's fight for pure and adequate water, told without color. Benefits to be derived. Bay counties and irrigation districts provided for in national grant. *San Francisco Chronicle* (1869-Current File). Retrieved from http://search.proquest.com/docview/365911607/abstract/F3507259CFB44C02PQ/4.

Minton, J. (2008). The old and the new: Evaluating existing and proposed dams in California. *Golden Gate University Environmental Law Journal, 2,* 96.

Modesto Irrigation District. (2020). Tuolumne River voluntary agreement Appendix A-6. Retrieved from https://www.untilthelastdrop.com/voluntary-agreement.

Mohanty, S. K., Valenca, R., Berger, A. W., Iris, K. M., Xiong, X., Saunders, T. M., & Tsang, D. C. W. (2018). Plenty of room for carbon on the ground: Potential applications of biochar for stormwater treatment. *Science of the Total Environment, 625,* 1644–1658.

Monroe, H. (1910). Shall the Hetch-Hetchy valley be saved for the nation? *Chicago Daily Tribune* (1872–1922). Retrieved from http://search.proquest.com/docview/173420193/abstract/769FA8A98204419CPQ/7.

Moyle, P. B., Herbold, B., Stevens, D. E., & Miller, L. W. (1992). Life history and status of delta smelt in the Sacramento-San Joaquin estuary, California. *Transactions of the American Fisheries Society, 121*(1), 67–77. doi: 10.1577/1548-8659(1992)121<0067:LHASOD>2.3.CO;2.

Muir, J. (1913). John Muir asks that their bill be not rushed through congress. *New York Times.* Retrieved from http://search.proquest.com/docview/97467783/abstract/769FA8A98204419CPQ/2.

Mukherjee, M., & Jensen, O. (2020). Making water reuse safe: A comparative analysis of the development of regulation and technology uptake in the US and Australia. *Safety Science, 121*(January), 5–14. doi: 10.1016/j.ssci.2019.08.039.

New York Times. (2002). Bring back Hetch Hetchy? Retrieved from http://search.proquest.com/docview/92185138/citation/C1C6FB0416F8438FPQ/2.

New York Times. (2012). Hetch Hetchy's past and future. Retrieved from http://search.proquest.com/docview/1705878441/citation/C1C6FB0416F8438FPQ/3.

Paradigm Environmental. (2021). *Concept for a multi-jurisdictional regional stormwater capture project.* Orange Memorial Park (City of South San Francisco).

Porse, E., & Pincetl, S. (2019). Effects of stormwater capture and use on urban Streamflows. *Water Resources Management, 33*(2), 713–723. doi: 10.1007/s11269-018-2134-y.

Public Policy Institute of California. (2016). Water for cities. Retrieved from https://www.ppic.org/wp-content/uploads/R_1016EH3R.pdf.

Ray, J. R., Shabtai, I. A., Teixidó, M., Mishael, Y. G., & Sedlak, D. L. (2019). Polymer-clay composite geomedia for sorptive removal of trace organic compounds and metals in urban stormwater. *Water Research, 157,* 454–462.

Richtel, M. (2015). Saving water by wagging fingers: [Science desk]. *New York Times, Late Edition (East Coast),* October 13, 2015. Retrieved from http://search.proquest.com/docview/1721534869/abstract/1B1BD2DBD10C4E17PQ/5.

Rodina, L., Baker, L. A., Galvin, M., Goldin, J., Harris, L. M., Manungufala, T., ... Ziervogel, G. (2017). Water, equity and resilience in Southern Africa: Future directions for research and practice. *Current Opinion in Environmental Sustainability,* Open issue, part II, *26–27*(June), 143–151. doi: 10.1016/j.cosust.2017.09.001.

Rohde, M. M., Reynolds, M., & Howard, J. (2020). Dynamic multibenefit solutions for global water challenges. *Conservation Science and Practice, 2*(1), e144. doi: 10.1111/csp2.144.

San Francisco Public Utilities Commission. (2014). *Water resources division annual report fiscal year 2013-2014.* San Francisco, CA.

San Francisco Public Utilities Commission. (2021a). SFPUC 10-year financial plan FY 2021–22 to FY 2030–31. Retrieved from https://sfpuc.sharefile.com/share/view/s3ec5be4f56 bd4d7eb94117c685e3b29f.

San Francisco Public Utilities Commission. (2021b). 2020 urban water management plan for the City and County of San Francisco public review draft. Retrieved from https://sfpuc.org/sites/default/files/documents/UWMP%20Public%20Review%20Draft %2004012021%20FINAL.pdf.

Sandkulla, N. (2020). *Statement from Nicole Sandkulla, chief executive officer, to the San Francisco public utilities commission (SFPUC), about San Francisco's obligations to provide a reliable supply of high quality Water at a fair price to its wholesale customers in alameda, San Mateo, and Santa Clara counties.* November 30, 2020. San Mateo, CA.

Santa Clara Valley Urban Runoff Pollution Prevention Program. (2015). Santa Clara Valley Urban Runoff Pollution Prevention Program 2014 program summary. Retrieved from https://cleancreeks.org/ArchiveCenter/ViewFile/Item/58.

Sapkota, M., Arora, M., Malano, H., Moglia, M., Sharma, A., George, B., & Pamminger, F. (2015). An overview of hybrid water supply systems in the context of urban water management: Challenges and opportunities. *Water, 7*(1), 153–174. doi: 10.3390/w7010153.

Sommer, L. (2014). Bay Area: Do you know where your water comes from? *KQED*. Retrieved from https://www.kqed.org/science/14623/bay-area-do-you-know-where-your-water-comes-from.

State Water Resources Control Board. (2016). Resolution no. 2016–0010. Retrieved from https://www.waterboards.ca.gov/board_decisions/adopted_orders/resolutions/2016/rs2016_0010.pdf.

State Water Resources Control Board. (2018). Resolution no. 2018–0059. Retrieved from https://www.waterboards.ca.gov/board_decisions/adopted_orders/resolutions/2018/rs2018_0059.pdf.

State Water Resources Control Board and California Environmental Protection Agency. (2018). Final substitute environmental document in support of potential changes to the water quality control plan for the San Francisco Bay-San Joaquin Delta estuary: San Joaquin River flows and Southern Delta water quality. State Claringhouse #2012122071. Retrieved from https://www.waterboards.ca.gov/waterrights/water_issues/programs/bay_delta/bay_delta_plan/water_quality_control_planning/2018_sed/docs/00_ES.pdf.

Superior Court of California. (1983). National Audubon Society v. Superior Court.

Taber, I. (1908). Looking up Hetch Hetchy valley from surprise point. *Sierra Club Bulletin, IV*(4), 211. Retrieved from https://vault.sierraclub.org/ca/hetchhetchy/looking_up_hh_valley_taber.html.

Teunis, P. F. M., Xu, M., Fleming, K. K., Yang, J., Moe, C. L., & LeChevallier, M. W. (2010). Enteric virus infection risk from intrusion of sewage into a drinking water distribution network. *Environmental Science and Technology, 44*(22), 8561–8566. doi: 10.1021/es101266k.

United Nations General Assembly. 2014. *"International decade for action 'Water for Life' 2005-2015."* https://www.un.org/waterforlifedecade/.

United States Environmental Protection Agency. (2019). National water reuse action plan draft. Retrieved from https://www.epa.gov/sites/production/files/2019-09/documents/water-reuse-action-plan-draft-2019.pdf.

Valley Water. (2019). *Water supply master plan 2040*. Retrieved from https://www.valley-water.org/sites/default/files/Water%20Supply%20Master%20Plan%202040_11.01.201 9_v2.pdf.

West, J. (2019). SB 332 (Hertzberg/Wiener) - Oppose unless amended. Retrieved from https://watereuse.org/wp-content/uploads/2017/03/SB-332-ltr-oppose-unless-amend-WRCA .pdf.

Willis, R. M., Stewart, R. A., Panuwatwanich, K., Williams, P. R., & Hollingsworth, A. L. (2011). Quantifying the influence of environmental and water conservation attitudes on household end use water consumption. *Journal of Environmental Management, 92*(8), 1996–2009. doi: 10.1016/j.jenvman.2011.03.023.

Xu, Q., Liu, R., Chen, Q., & Li, R. (2014). Review on water leakage control in distribution networks and the associated environmental benefits. *Journal of Environmental Sciences, 26*(5), 955–961. doi: 10.1016/S1001-0742(13)60569-0.

Zone 7 Water Agency. (2006). *Stream management master plan.* Alameda County, California.

Zone 7 Water Agency. (2014). *Preliminary lake use evaluation for the chain of lakes.*

6 An Eye to the Future – Implementing Climate-Resilient Approaches to Land and Water Conservation in the State of Minnesota

Beverly Rinke, Haley Golz, Kris Larson,
Nicholas Bancks, Virginia Breidenbach,
Wayne Ostlie, Net Phillips, and Ruurd Schoolderman

CONTENTS

DOI: 10.1201/9781003048701-6

6.1 INTRODUCTION

While climate change is impacting all corners of the United States, the State of Minnesota is experiencing some of the most dramatic impacts in the continental United States to date, according to the U.S. Environmental Protection Agency (U.S. EPA) (U.S. EPA 2016). Additionally, while Minnesota traditionally may grab the headlines for record cold, it is the second fastest warming state in the United States. The state has warmed 1–2 degrees in the last century, with the more rapid warming trends in the northern portion of the state. Average annual precipitation has increased in the Midwestern United States 5–10%, with rainfall on the four wettest days of the year having increased an average of 35%. This trend of increased rainfall and intensifying rain events is expected to increase, resulting in the heightened risk of floods (U.S. EPA 2016).

The changing climate has drastically increased the frequency of extreme weather events which has now placed Minnesota in the top 10 U.S. states with the highest weather-related insurance costs. Minnesota has experienced 23 billion U.S. dollars' worth of damage from weather events since 2000. As a comparison of how weather extremes have changed, a combined US$1.5 billion loss as a result of three weather events in 1998 was more than the total combined weather-related losses for the previous 40 years (Minnesota Public Radio (MPR) 2021).

These recent dramatic climate changes are not only impacting Minnesotans. The ranges of plants and animals are likely to shift as temperatures rise. For example, warmer weather will change the composition of Minnesota's forests. Warming could also harm ecosystems by changing the timing of natural processes such as migration, reproduction and flower blooming. Changes in these patterns can result in a misalignment of timing between co-dependent species, such as plants and pollinators, the impacts of which can cascade through an entire ecosystem (U.S. EPA 2016).

Fortunately, Minnesota is well positioned to take advantage of Natural Climate Solutions (The Nature Conservancy 2021) in order to minimize its contributions to greenhouse gases and help our natural systems be more resilient in the face of this change. According to a 2021 report from The Nature Conservancy, Minnesota could curb up to 16% of its 2005 level greenhouse emissions through Natural Climate Solutions (Ahlering et al. 2021).

Land conservation organizations such as the Minnesota Land Trust are well positioned to deploy strategies that allow natural systems to adapt to climate change while continuing to provide important ecosystem services such as flood protection and clean water, while mitigating greenhouse gas emissions. This chapter will provide concrete examples of how individual landowners and conservationists in Minnesota are taking action to help mitigate the impacts of climate change on our natural communities and the species on which they depend. While climate solutions will require that we change the trajectory of large, unwieldy systems like our energy and transportation sectors, the actions of *individual* people – with their own personal motivations and abilities – will provide the nudge necessary for larger change to take place.

In the following sections, you will read in four case studies how private landowners, who own 76% of Minnesota's landscape and more than 60% of the United States (Summit Post 2021)*, have been one of these important catalysts by voluntarily protecting or restoring their properties. This will in turn benefit specific natural communities both today and over the long haul by making them more resilient to a changing climate. And finally, you will also read about how nonprofit land trusts and other conservation entities are uniquely positioned to act on behalf of the public to address climate change. While land trusts have been a part of the American landscape since 1891, they have historically been focused on local or regional land use issues. Today, however, both their scope and importance have grown along with our understanding of the great potential our lands play in addressing climate change.

6.2   THE CONTEXT OF PRIVATE LAND CONSERVATION AND LAND TRUSTS

To better understand the case studies presented in this chapter, it is important to set the context for land conservation in Minnesota. Minnesota has long had a commitment to conservation. Several of the 20th-century strongest wilderness advocates called Minnesota home, such as Ernest Oberholtzer and Sigurd Olson. More recently, Minnesota residents voted to tax themselves to create a constitutional "Clean Water, Land and Legacy Amendment" which dedicates more than US$300 million a year to fund investments into clean water, parks, arts and wildlife habitat. This funding source provides strong support for the Land Trust and its partners to implement conservation and climate action strategies on a landscape level.

Minnesota Land Trust was born out of this conservation ethic in 1991, and the Land Trust is one of more than 1,500 land trusts across the country which share a common goal of protecting and restoring important lands within their service areas on behalf of the public.

---

* Data from is from the US Bureau of the Census, Statistical Abstract of the United States: 1991 (11th ed.) Washington, DC, 1991, p. 201.

While land trusts vary on their business models, many are dedicated to working with private landowners to voluntarily protect properties through conservation easements. Conservation easements are legally binding agreements between a landowner and land trust or government agency to give up certain rights (like the right to clear-cut or develop the property) in order to protect in perpetuity the property's important conservation values. Nationally, land trusts have preserved more than 56 million acres of land, more than twice the size of all our national parks combined.

In Minnesota, there are more than 12,000 individual conservation easements, making it one of the states with the greatest participation of private landowners in voluntary land conservation. These easements are "held" by either the federal government, state government, local government or a nonprofit land trust, typically the Minnesota Land Trust. This large-scale effort to secure conservation easements demonstrates just how important private landowners are in addressing many things Minnesotans hold dear, such as clean water, wildlife habitat and, more recently, climate change mitigation.

As a complement to its more traditional land protection tools, the Land Trust also employs ecosystem restoration in partnership with private landowners and government partners to proactively take action in promoting climate-resilient landscapes.

Minnesota Land Trust's involvement in climate change issues has increased along with the growing understanding of how much natural land solutions – which promote natural system resiliency, carbon mitigation, protection of ecosystem services and species migration in face of climate change – matter for climate action. Land trusts are uniquely positioned to facilitate protection and restoration work on private lands and serve as a vital link between private landowners and public resources that are available. These private landowners are often motivated to take action because of the strong connection they feel to the land and a desire to leave a lasting legacy. Minnesota Land Trust's commitment to make a lasting impact on a landscape level through what it calls "relentless incrementalism" positions it well to connect these private landowners to a larger coordinated effort and achieve more meaningful conservation.

This commitment to addressing climate change begins with conservation planning: where and how the Land Trust dedicates its resources. The conservation program focus-area planning process, which is described below in the first special study, serves as an important tool to define landscape-level strategies in which the Land Trust and its partners add an assessment of maximum climate action benefits to other conservation priorities such as wildlife habitat or clean water. This "resilient landscapes" planning process is predicated on data from The Nature Conservancy which highlights those natural communities which have the greatest potential to be resilient in face of the changing climate (The Nature Conservancy, Resilient Mapping Tool, 2021). The resulting plan serves as a catalyst for landscape-level conservation action informed specifically by a goal of building a resilient landscape in the face of a changing climate.

Climate science plays an important role in the Land Trust's strategic priorities. Not surprisingly, with Minnesota being the "Land of 10,000 Lakes" and encompassing the headwaters of the Mississippi River, a significant portion of the Land Trust's organizational effort focuses on protecting our state's water resources. Specifically, the Land Trust's conservation efforts target protecting our aquatic systems by protecting water quality, keeping cold lakes and streams cold and slowing down runoff by promoting natural systems that help store water on the land.

In addition, Minnesota's unique geography featuring transition zones between three major ecosystems (prairie, northern deciduous and boreal forests) places it on the frontline of moving climate zones. As noted previously, Minnesota is experiencing some of the most rapid and dramatic climate change within the continental United States, impacts of which are already seen on the ground in terms of species migration and decline and increased frequency of disturbance events such as windstorms and flooding.

6.3 FOUR SPECIAL STUDIES ILLUSTRATING CLIMATE-RESILIENT CONSERVATION ACTION

The following four special studies highlight how land trusts in particular and conservation organizations in general – through planning, partnerships and leadership focused on land protection and restoration – can serve as catalysts for climate action. They also demonstrate how important America's private landowners are as partners in addressing the impacts of climate change (Figure 6.1).

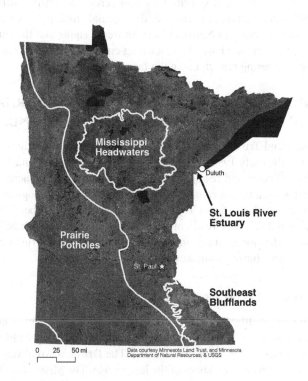

FIGURE 6.1   Special Study Program Areas.

The "Planning for Biodiversity Resilience" special study describes the conservation program area planning process to prioritize protection of cold-water fisheries of the Driftless Area's Blufflands with its heritage brook trout streams. This special study highlights use of a variety of data to develop priorities for a climate-resilient protection strategy.

The "Keeping Cold Water Lakes Cold" special study describes how Land Trust and its partner, Northern Waters Land Trust, leveraged science and relationships to mobilize major funding resources and ensure long-term protection of high-quality strategic tullibee refuge lakes that can withstand projected climate change with adequate watershed protections in place.

The "Resilient Agricultural Landscapes: Restoring, Connecting and Keeping Water and Soil on the Land" Prairie Pothole region special study highlights the importance of public–private landowner and nongovernmental organization (NGO) partnerships aimed at protecting and stitching back together resilient systems in an ecoregion that has seen tremendous conversion of its original habitat to agricultural use.

Finally, the "Coastal Habitat Protection" special study demonstrates the leadership role a land trust can play in mobilizing funds and bringing together a complex multi-agency partnership. The St. Louis River, the largest estuary system within Lake Superior, offers a special study for the role conservation organizations can play in climate action in a dynamic system with commercial waterfront, urban development and recreational activity. In addition, this special study provides an important cultural dimension relating to Native American treaty rights and the importance of climate action to protect the basis for sustaining culturally relevant resources (hunting, fishing and gathering resources) in the face of climate change.

6.4   PLANNING FOR BIODIVERSITY RESILIENCY – HERITAGE BROOK TROUT STREAMS MINNESOTA BLUFFLANDS

The Minnesota Land Trust has been actively working in the Blufflands of Southeast Minnesota since the early 1990s. To date across the region, the Land Trust has protected over 10,000 acres and restored 200 acres with another 200 acres in the pipeline in a program area that is roughly 700,000 acres in size. Recognizing the scale of the challenge ahead, the Land Trust undertook a planning process to prioritize, target and develop its climate-resilient strategies across the Blufflands. This special study lays out the development and implementation of the Land Trust's conservation strategies while also sharing some key insights from the process.

6.4.1   BLUFFLANDS LANDSCAPE CONTEXT

Southeast Minnesota's Blufflands is part of a larger geological region known as the Driftless Area, a roughly 24,000 square mile swath of land lying within portions of four Midwestern U.S. states (MN, WI, IA, IL). The Driftless Area was left unglaciated during the most recent advances of the last ice age. The Mississippi River and its major tributaries in southeastern Minnesota – the Cannon, Zumbro, Whitewater and

Root Rivers – carved a scenic karst landscape characterized by a deeply dissected plateau of valleys, high bluffs and numerous cold-water streams.

This spectacular terrain was formed due in part to the absence of glacial drift and also due to the unique karst topography of the area, created by the interaction of water with the underlying soluble limestone and dolomite bedrock. This interplay of water and rock manifests itself in the many sinkholes, caves, disappearing and underground streams, springs and seeps and cold-water streams characterizing the region.

The Blufflands comprises the most biodiverse landscape in Minnesota – known to being home to 159 Species in Greatest Conservation Need (SGCN), more than anywhere else in the state. In addition, the area is a critical flyway for migratory birds (MN DNR 2006), known as the "Mississippi Flyway." Groundwater-fed cold-water streams are common, supporting important aquatic habitat and providing plentiful recreational opportunities. Within these cold-water streams exists one of the iconic species to the region, the brook trout (*Salvelinus fontinalis*). Unlike brown (*Salmo trutta*) and rainbow trout (*Oncorhynchus mykiss*), both of which have been introduced to the streams of the region, the brook trout is the only one native to the Blufflands. Beautifully colored, with a dark background of blacks, purples and greens highlighted by red spots along its side, brook trout can be found in cold, clean waters of the Blufflands, especially near headwaters.

6.4.2 BLUFFLANDS'S CLIMATE THREAT REQUIRES BROAD WATERSHED-BASED PRIVATE LANDS PROTECTION

Southeast Minnesota faces significant conservation threats posed by climate change. Recent trends point to increasingly warmer and wetter climatic conditions (Chismar 2015). In the last 20 years alone, "mega-rain" events (rainfall events that cover >1,000 square miles and drop >6 inches of precipitation) have increased twofold relative to the previous 30 years. The City of Hokah, located in Houston County, set the state record in 2012 receiving 15 inches of rain in a 24-hour period (MN DNR 2021). Erratic and intense storm surges, like mega-rain events, negatively impact features like cold-water streams through the large amounts of water they bring into systems which increase nutrient and sediment loading while subsequently reducing water temperature, oxygen levels and clarity.

Stable and sufficient baseflow is important to maintaining cold-water system and aquatic communities. Recent research from the Minnesota Department of Natural Resources (MN DNR) suggests that consistent baseflow from groundwater springs can provide a level of resilience to these cold-water systems (Tipping et al. 2019). To achieve this, protection of riparian areas immediately adjacent to these streams alone is not enough. Cold-water systems are groundwater dependent and the porous karst topography creates myriad connections and influence points throughout an entire stream's watershed. This requires a broader watershed protection approach.

Ensuring resilience of the system requires extending protection and restoration efforts to areas beyond the riparian corridor to a larger portion of the watershed. Outside the Mississippi River corridor, much of this area is privately owned, necessitating a strong private lands strategy for the protection of this important Minnesota landscape and its cold-water resources. Of special concern are those areas most at risk of conversion to development, the bluff shoulders and toes with thin erodible soils and rapid connections to groundwater. Protection of bluff tops near or at the headwaters of these cold streams is equally vital because these are typically areas of intensive agricultural use where sink holes directly link surface to groundwater sources, thus influencing quality and base flow.

6.4.3 BLUFFLANDS CONSERVATION PRIORITIES, PLANNING AND PARTNERSHIPS

The understanding of the necessary conditions to support healthy and resilient cold-water systems informed the development of strategies and targeted conservation action across the Blufflands. With limited resources to implement conservation actions, it was clear that science-based conservation planning and strategic deployment of available resources was necessary to efficiently leverage the limited funding and staff capacity. The planning efforts which underpinned the conservation actions utilized existing data and reports on watershed conditions and prioritized conservation efforts to the most impactful areas across Southeast Minnesota's Blufflands region.

In this process, the Land Trust recognized the value and necessity of strategic partnerships to aid in tackling the conservation priorities in order to achieve desired outcomes. Several conservation organizations have been active in the area. These included private nonprofits such as The Nature Conservancy (TNC), Trout Unlimited and The Trust for Public Land, as well as public entities such as the MN DNR, U.S. Fish and Wildlife Service and local Soil and Water Conservation Districts. These organizations formed a robust conservation partnership focusing specifically on cold-water heritage brook trout stream protection. This partnership recognized that the strengths and capacity each partner brings are instrumental to achieving any lasting conservation impact.

The brook trout is an apt indicator which was used to assess the resiliency and health of cold-water systems. Because of their dependence on cold, clean water, brook trout are highly susceptible to low oxygen levels and warm water temperatures caused by excess nutrient and sediment loading to streams and climate change. The presence of "Heritage Brook Trout" (HBT) populations, which are those with genetic strains unique only to the region pre-dating any modern stocking efforts, can indicate which watersheds across the Southeast Blufflands might be most inclined to function as future cold-water refugia given that they currently support viable HBT populations. Recognizing where these HBT populations exist, and in some instances are thriving, was a critical first step in assessing where to target conservation action that would have a significant impact in maintaining resilient systems in the presence of changing climate.

Research led by several fisheries biologists from the MN DNR provided extensive information regarding the distribution, genetics and abundance of brook trout across the Driftless Area of Minnesota. Once thought to be nearly extirpated from southern Minnesota, HBT are now present in roughly 20% of the cold-water streams of the Southeast Blufflands (Hoxmeier et al. 2015) (Figure 6.2).

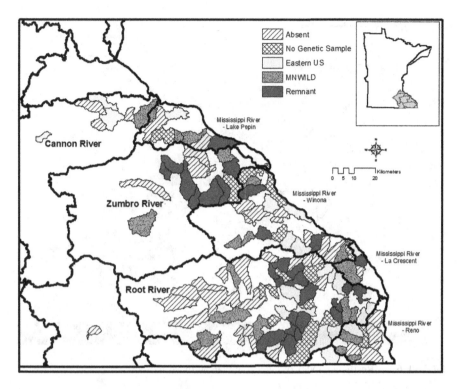

FIGURE 6.2 Distribution and genetic strains of brook trout in southeastern Minnesota delineated by major watersheds.

Natural resource managers and planners in Minnesota have a robust and extensive array of conservation data and planning tools available. This includes spatial information regarding HBT population location and abundance. In addition to this, several other analyses were taken into consideration to prioritize watersheds for protection and restoration efforts. The MN DNR's *Watershed Health Assessment Framework* (WHAF) and TNC's *Resilient and Connected Landscapes* report served as the foundational components to identify resilient watersheds. The WHAF provides an overall health score for watersheds across Minnesota, considering multiple criteria as part its scoring process. Key metrics in our resilience analysis from this dataset included spring density, areas of high proportion of perennial cover, minimal wetland loss and impervious cover, aquatic Index of Biotic Integrity scores and aquatic and riparian connectivity. The TNC *Resilient and Connected Landscapes* report provided a comprehensive analysis to identify resilient sites that are more likely to maintain biodiversity and ecological function in the face of changing climate conditions (Figure 6.3).

Combined, these powerful analytic tools were used to guide and direct conservation activity. These two datasets were integrated and overlaid with HBT distribution and abundance. This provided a prioritization of watersheds across the Southeast

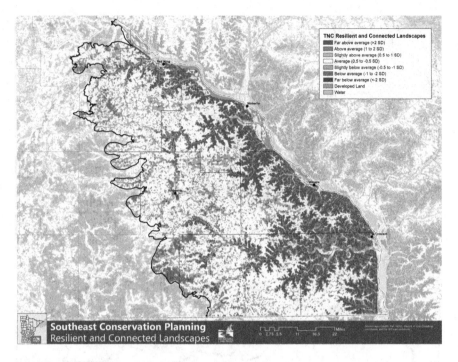

FIGURE 6.3 Resiliency index of Paleozoic Plateau in Minnesota.

Blufflands where land protection could have the most significant impact on watershed health and climate resiliency (Figure 6.4).

The overlay of the three datasets (HBT density and abundance, WHAF, Resilient and Connected Landscapes) resulted in the identification of six priority watersheds, ranging from 60 to 400 square miles in size, where significant populations of HBT exist with high watershed health and connectivity. These watersheds were determined to possess certain physical characteristics, in enough quality and condition, to support ecological function and species diversity as the climate changes. Collectively, these six watersheds were selected as a focused target area to achieve measurable conservation goals. These select watersheds offer the best hope of maintaining the cold-water systems of Southeast Minnesota as its climate becomes wetter and warmer.

Within these six watersheds, 97% of the land base is in private ownership. This presents a major need and opportunity for the Land Trust to engage with landowners and deploy conservation easements and restoration as land protection tools.

6.4.4   Plan Implementation

Minnesota's Outdoor Heritage Fund (OHF) – one of three primary funds created through the passage of the Clean Water, Land and Legacy Amendment in 2008 – plays a critical role in providing the financial resources crucial for undertaking land

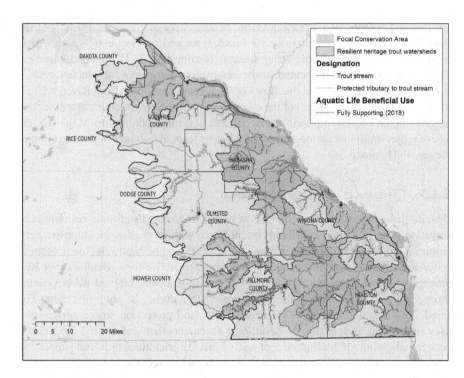

FIGURE 6.4 Heritage Brook Trout Watersheds in Southeast Minnesota.

protection and restoration in the Blufflands region. The Land Trust, through its partnership in the Blufflands, has secured millions of dollars for conservation in Southeast Minnesota from the OHF since it came into existence. The Land trust's role in this partnership effort is targeted easement acquisition in conjunction with habitat restoration, focusing on parcels next to existing protected habitat complexes, along habitat corridors, adjacent to riparian areas, and in headwaters.

This targeted and strategic approach to priority watershed protection has resulted in some significant success in protecting larger protected land complexes. As part of a larger conservation partnership, the Land Trust has been working to protect and restore natural lands across Southeast Minnesota. TNC and Trust for Public Land (TPL) have focused protection efforts on fee acquisition, eventually conveying lands they purchased to the MN DNR to become part of the public lands base. Through these efforts, a 900-acre WMA was acquired in 2018, and several adjoining parcels are in process to be added to the 2016 WMA acquisition which will grow the existing protected habitat complex.

Between 2019 and 2021, in the South Fork Root River, the Land Trust has already acquired two conservation easements, both in strategic locations on the landscape. The first easement acquisition is adjacent to a 1,000-acre State of Minnesota Wildlife Management Area (WMA), acquired by and subsequently conveyed to the State of Minnesota in 2016 by TNC. The Land trust easement, 416 acres in size, protects the forested uplands and a portion of the riparian area along the south fork of the Root River, an important corridor for brook trout and other aquatic species linking to numerous smaller cold-water streams feeding into it. The second easement acquisition (203 acres) is located in the headwaters catchment for Maple Creek, a state-designated trout stream that flows into the south fork of the Root River, approximately 2.5 miles downstream of the easement. In this case, the Land Trust easement protects upland forest, riparian corridors and several tributaries to Maple Creek. These projects are just the beginning of a longer term investment in conservation in the region (Figure 6.5).

6.4.5 CHALLENGES AND LESSONS LEARNED

Despite these successes, implementation of the Land Trust's climate-resilient conservation strategies in the Blufflands has not been without challenges. Intensive agricultural land use is still present in many of the bluff top and headwater areas of these watersheds. These areas are generally the most productive for agriculture, yet land management associated with this often has negative impacts on cold-water systems through increased nutrient loading and water surge during heavy rain events. The Land Trust continues to adapt its approach to land protection where agricultural uses are present. It does this by developing easements that seek to find the balance between allowing for continuation of some limited agricultural uses and supporting land management practices that reduce impacts and protect the conservation values of easement parcels.

It is important to recognize that partnership development and maintaining relationships with private landowners is a time-intensive effort. Conservation requires a commitment to the "long game" where the arc of protection and restoration can take

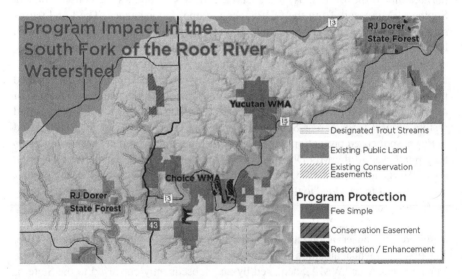

FIGURE 6.5 Protection impact in the South Fork Root River watershed.

years before desired results come to fruition. This long-term time investment contrasted with the rapid and dynamic shifts posed by climate change and development pressures poses a challenge to conservation efforts and requires decisive action now.

6.5 KEEPING COLD WATER COLD – MISSISSIPPI HEADWATERS CISCO REFUGE LAKES

6.5.1 MISSISSIPPI HEADWATERS LANDSCAPE SETTING AND THREATS

The Mississippi Headwaters in north-central Minnesota is home to some of the nation's most pristine freshwater lakes. This approximately 9,100 square miles area offers a landscape of mixed-deciduous and pine forests which provide a scenic backdrop to the area's abundant lakes. These lakes, treasured as an iconic recreational asset by millions of Minnesotans, are a state priority for conservation.

The Mississippi Headwaters region boasts a relatively pristine forested landscape with an existing high level of protected land area. Currently, approximately 68% of the total land base in the Land Trust's Mississippi Headwaters Program Area has some form of permanent protection, either as public land (in the form of state and federal forests, state wildlife management areas and others) or as private lands protected through conservation easements. However, despite this already high level of protection and relatively low land disturbance, these lakes are under increasing pressure of degradation in part as a result of warming temperatures and shoreland development pressure.

The deep cold-water lakes, which are a protection priority in the program area, are oligotrophic (nutrient-poor) phosphorous-limited lakes. This makes them especially sensitive to increased nutrient runoff resulting from (shoreland) development and land disturbance such as logging and agriculture. The area's status as a popular vacation and recreational destination has contributed to development pressures, especially on shoreland. As the prime shoreland building sites have been developed, more marginal and sensitive shorelands are experiencing increased development pressure. Development of these more marginal sites is a concern for both nutrient runoff and habitat impacts as these are often ecologically valuable areas which contain wetlands and sensitive shoreline that are important in protecting water quality and provide important habitat.

Climate change is amplifying the negative impacts from shoreland development. State of Minnesota monitoring data has found that water temperatures in the Minnesota Central Lakes region are warming at a faster rate than the rest of the state (see Figure 6.6; Olmanson 2021). This warming trend is believed to be a contributing driver for decreasing water clarity as a result of increased summer algae blooms.

The combination of increasing lake water temperatures and more frequent algae blooms threaten the cold-water-dependent fisheries of these deep cold lakes. The tullibee or cisco (*Coregonus artedi*), a small silvery fish in the family of Salmonidae, is the main food source for prized game fish such as walleye and iconic birds such as Minnesota's state bird, the common loon. Cisco are very temperature sensitive, requiring well oxygenated water with temperatures below 17°C (Jacobson et al. 2010).

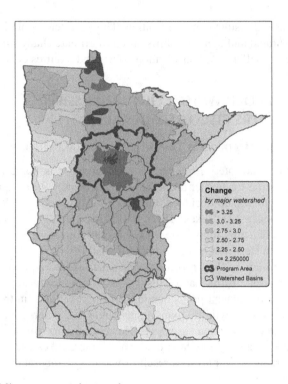

FIGURE 6.6 Minnesota warming trends.

As the water column warms up in the lake over the summer and decaying algae sinks to the bottom and consumes oxygen, these fish get "squeezed" in a smaller band of water that meets their both temperature and oxygen requirements. This squeeze can result in these important fish disappearing from the lakes. With that, the entire food web can unravel impacting a host of other fish as well as birds that depend on these fishes as a food source.

6.6 SETTING PRIORITIES IN PROTECTING THE LAND OF 10,000 LAKES

The Minnesota DNR has identified 68 primary cold-water refuge lakes in Minnesota (Minnesota DNR 2021). Of these, 38 are located in the Land Trust's Mississippi Headwaters Program Area (Figure 6.7). These refuge lakes are deep lakes located in relatively intact watersheds that, when adequately protected, can continue to support a cold-water fishery in the face of climate change. While large areas of forested land in these refuge lake watersheds are under public ownership, a considerable amount of land is also owned by private individuals. These parcels are increasingly being "split up," sold and developed.

Modeling suggests that total phosphorus concentrations remain near natural background levels when less than 25% of a refuge lake's watershed is disturbed

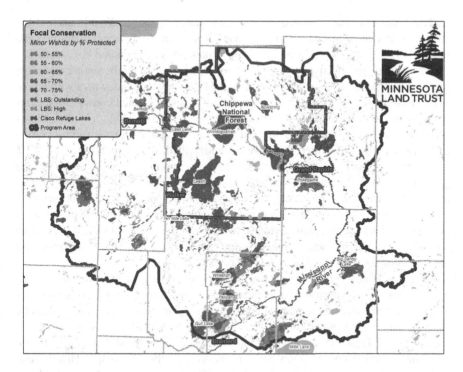

FIGURE 6.7   Mississippi headwaters priority lakes for protection.

(Jacobson et al. 2013). Protection of private lands play a critical role in order to achieve a 75% watershed protection rate. In addition, shorelands, which are of extra critical importance in protecting water quality and provide important near shore fish habitat, are mostly privately owned. For this reason, conservation easements are an important tool to work with these private landowners to protect their lands.

6.6.1   Strong Partnerships and Committed Private Landowners Needed for Success

Protecting the watersheds of 38 tullibee refuge lake is an enormous undertaking. Fortunately, thanks to a combination of research and a strong partnership between government and private nonprofit conservation organizations, a concerted effort is underway to move these lakes to the goal of 75% watershed protection.

Within this constellation of conservation partners, Land Trust and its partner Northern Waters Land Trust (NWLT), have together carved out their own unique niche to contribute to the 75% watershed protection goal. NWLT is a regional land trust with a specific focus on the Mississippi headwaters. They play a key role in outreach to identify willing landowners that are interested in placing an easement on their property. They do this through community outreach and their local network and close contacts with the lake associations that are organized around many of the target lakes.

This landowner outreach is done in a very strategic manner. Within each of the target watersheds, every parcel greater than 20 acres in size has been identified and ranked (Figure 6.8). This prioritization map takes into consideration size, wetlands, conservation value ranking and proximity to other protected lands, among other factors. The scoring provides a list of priority lands, associated landowners with whom to connect and ultimately guidance as to where to invest limited conservation easement dollars for the greatest conservation impact.

Once landowners interested in land protection – either through conservation easement or sale of fee interest – have been identified, the Land Trust and NWLT pursue parallel tracks of action in pursuit of protecting these lands. The Minnesota Land Trust works with landowners interested in retaining their fee interest in the land while permanently protecting the associated conservation values. Conservation easements allow for flexibility to accommodate landowner needs while protecting a property's inherent conservation values. In addition, the easement valuation is based on an appraisal of the actual market value of the easement. This is an important consideration for protecting high-value shorelands.

Northern Waters Land Trust negotiates the fee value purchase or donation of high-quality shoreland properties which are then transferred to a government entity to manage long term as a public-protected area (e.g., State Wildlife or Aquatic Management Area, County Forest).

This Land Trust–NWLT partnership is buoyed by the complementary action of numerous other conservation partners working in the Mississippi Headwaters, including private NGOs such as The Nature Conservancy, Trust for Public Land

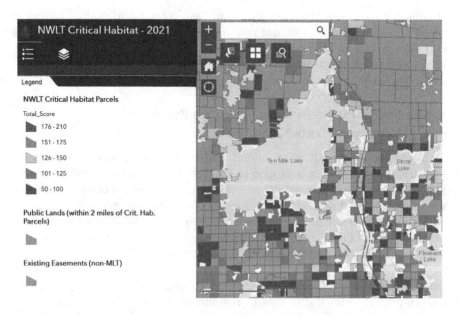

FIGURE 6.8 Parcel prioritization map.

and The Conservation Fund. In addition, local government agency partners include the Mississippi Headwaters Board, County Soil and Water Conservation Districts and public land management agencies, including the US Forest Service, U.S. Fish and Wildlife Service, Minnesota Board of Soil and Water Resources, MN DNR and County Forest Departments. These conservation partners meet and coordinate on a quarterly basis through the "North Central Conservation Round Table," which is facilitated by The Nature Conservancy.

Each partner within this constellation brings different tools for private land protection to the table. Depending on the property characteristics and landowners' needs and preferences, a different protection strategy may be selected. For example, Soil and Water Conservation Districts and Mississippi Headwaters Board protect lands through the Reinvest in Minnesota (RIM) easement program. These conservation easements offer an attractive and streamlined process to protect large blocks of typically non-riparian land. The payment formula and easement terms are non-negotiable. This setup makes it an attractive tool for protection of large tracts of forest lands, or marginal farmlands.

On the other end of the protection spectrum, an organization such as The Conservation Fund brings its own unique financing instrument by raising money through green bonds and has a focus on purchasing large tracts of industrial forest that have come on the market in Minnesota (The Conservation Fund 2020). After placing protections on it, the Fund will be transferring these properties to a long-term management entity such as a county land department, tribal entities or the MN DNR.

Private conservation-minded landowners who are interested in pursuing a permanent conservation easement on their property play a critical part in all of this. Often the multi-generational connection that people have to the land drives their desire to protect these special places for the next generations, like the Gouze family easement demonstrates.

The Gouze family easement is located on Lake George and Washburn Lake in Cass County, Minnesota. Beginning in the early 1990s, the Gouze family focused on buying contiguous tracts of sensitive land that had a clear impact on the lake's ecosystem. What drove them was the promise of protecting nesting sites for loons and the subtle beauty of the northern forest. Today it adds up to three miles of shoreline and 236 acres of forest that are all protected forever through a conservation easement with the Land Trust.

A perfect fit for their conservation goals, the Gouze's decided to put their accumulated property into a conservation easement with the Land Trust. "My mom was very excited about the prospect of protecting it in perpetuity," said Katie, one of the family members. "Creating the legacy was very important to them. They want this work to be bigger than this generation."

For people like Steve and Candace Gouze, their grandchildren will be the next generation in their family to spend summers catching walleyes and watching baby loons grow up in the protected bays and wetlands near their cabin. "I don't know what it's going to be like for them when they are adults," Candace said of her grandkids. "But this is one way that we can try to make sure that we will all be better off."

6.7 MOVING THE PROTECTION-LEVEL NEEDLE BY BUILDING ON SUCCESS

The simple, clear and consistent message by all partners – that we work in concert to protect 75% of the watershed to protect these refuge lakes and its iconic fisheries – has been effective in getting support at the local and state levels, and from private landowners who actively support conservation. A focused and coordinated approach on a sub-watershed and landscape scale has been very successful in working toward long-term climate resiliency of high-quality natural resources.

So, is this partnership moving the needle? The answer is yes! Protection is a long game and a team effort of relentless incrementalism. Over the past five years, together with our partners, we have already been able to move multiple watersheds (Ten Mile, Washburn, Girl and Woman Lakes) over the 75% protection level and made significant progress on many more. This proves that long-term protection is possible and that private landowners are interested in playing an active role by committing to long-term conservation of their lands.

Finally, success builds success. The partnership's ability to demonstrate success has resulted in continued strong support in terms of both funding and political support from state and local elected officials. Landowners who have successfully completed a conservation easement have become powerful advocates that help bring in new interested landowners, which keeps the momentum going.

6.8 RESTORING ECOSYSTEM SERVICES – KEEPING WATER ON THE LAND

The Prairie Pothole region of Minnesota, a captivating, rugged postglacial landscape created by the stagnation of massive glaciers during Minnesota's last glacial episode, is part of a more expansive manifestation of this system that sweeps across three Canadian provinces and five U.S. state. As the glaciers retreated over 10,000 years ago, millions of shallow depressions were left in the earth. These round depressions often fill with snowmelt and rainfall in the spring, creating valuable seasonal wetlands that support rich plant and animal life. Prairie pothole wetlands and grasslands together serve as natural sponges that hold excess water, reducing flooding and recharging groundwater systems, as well as offering important habitat for game species and a diverse group of nongame species, including pollinators, rare plants, insects and migratory birds.

The highly productive freshwater wetlands and surrounding grasslands are critically important to birds. The Prairie Pothole region of Western Minnesota is often referred to as North America's "duck factory" because it is estimated that about one-third of the continent's waterfowl breeding population nest within it, many of which spend the winter months in the coastal marshes along the Gulf Coast.

The region provides breeding habitat for a wide diversity of wetland- and grassland-dependent birds and provides stopover habitat for significant numbers of spring and fall avian migrants and native pollinators such as the monarch butterfly. Many charismatic neotropical migrants, such as the bobolink, dickcissel and raptors, rely on the shallow-water and prairie habitats.

The Prairie Pothole region is also home to a large variety of Minnesota Species in Greatest Conservation Need (SGCN) such as trumpeter swans, western prairie chicken, western grebe, American woodcock, Forster's tern and common loon, as identified by the DNR in *Minnesota's Wildlife Action Plan, 2015–2025*.

6.9 THE PRAIRIE POTHOLE REGION'S AGRICULTURAL IMPORTANCE RESULTED IN FRAGMENTED AND DEGRADED HABITAT

Once a vast grassland system, the Prairie Pothole region is now an agrarian system dominated by cropland. Before settlement by people of European descent, the pothole region was dominated by maple-basswood forests interspersed with oak savanna, tallgrass prairie and oak forest. Now across this region of Minnesota, more than 90% of wetlands have been drained and only about 1% of native prairie remains on the landscape. These alterations have drastically reduced the ability of the Prairie Pothole region to meet the needs of the vast array of birds that depend upon it sometime during their annual cycle. Wetland drainage and agricultural development have removed hundreds of thousands of acres of nesting, brood-rearing and foraging habitat for waterfowl as well as grassland gamebirds and songbirds (Minnesota Prairie Plan, 2nd Edition 2018).

These alterations have also directly impacted the ability of the landscape to hold water on the land. The agricultural dominance of this region also means that as water flushes directly into waterways, it negatively impacts water quality with increased sediment loads and agricultural nutrient and pesticide runoff.

6.10 CLIMATE THREATS AND OPPORTUNITIES FOR WIN-WIN RESTORATION BY KEEPING WATER ON THE LAND

The Prairie Pothole region, already threatened by habitat fragmentation and degradation, is facing serious impacts from climate change. Signs of climate change can be overlooked because weather extremes have been an inherent part of life on the prairie. Seasonally, there can be extreme temperature changes and high winds can lead to very dry conditions. Seasonal wetlands are typically wet for a few weeks after snowmelt or a heavy rainstorm, but are usually dry by midsummer.

Climate change may shift these seasonal patterns to a tipping point and dramatically reduce the habitat provided by prairie potholes during critical stages of the wildlife's life cycle. Climate models predict that the region will see greater extremes between periods of wet and draught conditions (U.S. EPA 2016).

During drought periods many prairie potholes are expected to dry up more frequently or earlier in the spring. This will lead to a decline in breeding habitat for waterfowl, and declines in other wetland species as well. However, predicted increases in large precipitation events may also lead to increased flooding as well as sediment, nutrient and chemical loading in our waterways.

Despite the loss of habitat and threat of climate change, millions of seasonal wetlands and large tracts of native prairie remain to make the Prairie Pothole region "one of the most altered — yet one of the most important — migratory bird habitats in the Western Hemisphere," as noted in the Prairie Pothole Joint Venture's 2017 Implementation Plan (Brice, J. C., Carrlson, K. M., Fields, S. P., et al. 2017). Therefore, to build resilience and to adapt to both extremes, it will be imperative to restore this landscape and increase its capacity to retain water on the land.

Efforts to increase habitat and water holding capacity in the Prairie Pothole region will be uniquely dependent on private landowners. Unsurprisingly, given the heavy concentration of agriculture in this region, almost 90% of the prairie potholes in Minnesota are privately owned. However, the fundamental water needs of local agrarian communities to produce crops and raise livestock and the needs for water and wetlands for wildlife are not incompatible. Both will benefit from habitat restoration projects that provide critical resources to wildlife while also working to remake the Prairie Pothole region into the "sponge" it once was. The same water retained on the landscape to be enjoyed by the birds will not cause flooding damage to farmlands. This type of conservation work is especially powerful. By joining partners in addressing factors that impact the interests of both communities, conservation can be advanced.

6.11   THE LAND TRUST AND USFWS PARTNERSHIP OFFERS A VEHICLE FOR PRIVATE LANDOWNERS TO PROTECT AND RESOURCES RESTORE THEIR LAND

Minimizing impacts of climate change requires programs to secure conservation easements on private lands, to discourage further draining or plowing of the remaining pothole wetlands and to restore wetlands and grasslands wherever possible. Luckily, there are many private landowners who deeply value their lands and want to see them preserved in perpetuity. Most landowners have profound personal connections to their properties borne out of countless hours working on or enjoying the land and value them intrinsically. Many of these landowners simply lack the resources or knowledge to protect, preserve or restore their properties.

Conservation easements offer a tailored approach that appeals to many landowners. The conservation easement allows landowners to own and manage their land while having access to professional guidance on protecting and improving their habitat. As part of this process, a habitat management plan is developed by an ecologist to clearly document existing plant communities on the property, invasive species or pathogens, and provide guidance on management or restoration strategies.

If in the process of establishing an easement it is determined that a property could benefit from more intensive management practices such as full-scale habitat restoration, the Land Trust also works with the landowners to secure the resources and guidance they need to undertake such a project.

Habitat restoration can take many forms, from removing invasive species to reconstructing a prairie from scratch in an old agricultural field. The unifying theme in "restoration" is that it is often a greater lift, in terms of both time and resources, compared to less intensive management practices, but with higher rewards. Some landowners are passionate enough to undertake restoration activities on their own, even if it means doing them incrementally over many years. However, the Land Trust often plays a crucial role at this juncture, connecting interested landowners who would otherwise be unable to undertake habitat restoration projects to the resources and expertise needed to restore wetlands and prairies.

An important resource that the Land Trust brings to landowners is connections with other agencies and their unique skills and opportunities. In the Prairie Pothole region, the Land Trust serves as a bridge to a particularly productive relationship with the U.S. Fish and Wildlife Service (USFWS). Combining our different strengths, the Land Trust and the USFWS have worked together to restore thousands of acres of high-quality, privately owned prairie pothole habitat each year. The USFWS's Partners for Fish and Wildlife program has a large portfolio of landowners interested in restoration and a 30-year history of working with landowners to advance conservation and restoration in the region. The Land Trust brings to the partnership an ability to attract landowners interested in more tailored conservation easements, as well as increased financial resources to accomplish restorations across many acres of both FWS and Land Trust easements. By bringing together each organization's strengths and unique relationships with landowners, this partnership is able to achieve major gains toward landscape-level resilience in the Prairie Pothole region.

6.11.1 PRIVATE LANDOWNERS ARE CRITICAL FOR LONG-TERM LAND STEWARDSHIP

Land conservation in this context is a three-way partnership between the Land Trust, USFWS and each individual landowner. As with Land Trust's conservation easements, each restoration undertaken by the Land Trust is tailored to meet the specific needs and interests of the private landowners and their property. The Pat Douglas case study in the following section illustrates this.

6.11.2 PAT DOUGLASS SPECIAL STUDY

Pat Douglass, originally from Southwestern Minnesota, attended college at the University of Minnesota at Morris, where he spent considerable time on the land in the Prairie Pothole region. During that time, he was an avid waterfowl hunter, and his senior thesis focused on waterfowl migration. From an early age, he was fascinated by birds and had always looked for ways to observe and understand more about them. When presented with the opportunity after college to purchase a piece of marginal farmland in the region he had come to love, Pat jumped at the chance. From the beginning, his objective was to restore this land to the postglacial landscape fresh in his mind from his college Geology class.

Pat was eventually able to buy a 640-acre section of land, deep into the historic tallgrass prairie. Pat worked with the Land Trust and the USFWS, as well as the

Reinvest in Minnesota (RIM) program, to place easements on the property and restore marginal farmland to a mixture of grasslands and wetlands. This open, rolling landscape now boasts a 150-acre open water wetland and more than 30 smaller wetlands surrounded by highly diverse prairie. This transformation is particularly important given the landscape context. Although Pat's property creates a small complex of natural habitat with nearby public land to the north and east, the west boundary of his property marks the beginning of a large area heavily influenced by agriculture (Figure 6.9).

The restoration of this property has had a profound impact on the local watershed. Early on, Pat could see the potential to bring back wetlands on his property. After big rain events, even when the property still held an extensive ditch system, he could clearly see the traces of historic wetlands. The wetlands were drained of water-holding capacity by building ditches leading out of each wetland; however, after installing plugs in these ditches, the land once again holds a significant amount of water during big rain events. Pat is expressly thanked by his neighbors, whose farms benefit from the decreased amount of saturation and flooding on their adjacent properties. The restoration activities have also transformed the property into a haven for birds (Figure 6.10).

Pat spends most of his time on the property either birdwatching or hunting. He has diligently tracked the species visiting his property over the years and has seen a

FIGURE 6.9 Visualizing relentless incrementalism.

FIGURE 6.10 Douglass wetland: a haven for birds.

FIGURE 6.11 Douglass prairie at sunrise.

marked increase in waterfowl and other birds on his property. Every year he watches the migration of ducks and prairie chickens on their distinctive leks (Figure 6.11).

Private landowners, like Pat Douglass, play an important role in the long-term stewardship of the land after the restoration is completed. These private landowners are excellent stewards who intimately know and care for their land. They conduct focused intensive monitoring and management and are able to carry out time-intensive treatments that would be cost prohibitive on public lands, such as non-chemical invasive species control or hand planting tree seedlings.

Uniting individual efforts like these, the Land Trust and the Partners for Fish and Wildlife program have restored 1,300 acres to date in the Prairie Pothole region and are in the process of restoring another 4,000 acres, and have another 1,900 in the queue, with more interest accumulating every day. The cumulative effect of these individual projects is having a huge impact in the Prairie Pothole region and across Minnesota. This partnership is actively stitching together a matrix of protected land which restores and connects large tracts of intact habitat. This in turn offers travel corridors for species to move and adjust to climate change and builds a landscape that protects watersheds.

This relentless incrementalism in the face of habitat degradation and fragmentation, and in the face of climate change, is only possible by aligning interests across organizations and individuals. The long-term relationships that the Land Trust maintains with our landowners is essential to help continually update management strategies and ensure that we are maintaining healthy ecosystems for people, water and wildlife alike.

6.12   COASTAL HABITAT PROTECTION IN THE ST. LOUIS RIVER ESTUARY

The St. Louis River Estuary is the largest freshwater estuary in North America and the largest tributary to Lake Superior in the United States. This unique 12,000-acre wetland complex, where water from the river mixes with the water from Lake Superior, is located between the cities of Duluth, MN, and Superior, WI. This mixing produces an incredible amount of biological productivity, and numerous species of fish, birds and other wildlife depend on the estuary for their survival.

The estuary is an important migratory bird corridor with exceptional species diversity. Over 230 species of birds, representing 75% of the bird species in Minnesota, use the estuary, while only about 75 of these are known or presumed to breed in the estuary (Audubon 2021).

Historically, the estuary teemed with wild rice or manoomin, as it's known to the Ojibwe people who have inhabited the region since long before it was settled by Europeans, as well as lake sturgeon or namé. Spirit Island located in Spirit Lake in the middle of the estuary was the location of the Sixth Stopping Place of the great migration westward of the Ojibwe people. It is a place of great cultural importance, where the "food that grows on the water" (manoomin) was first encountered (Schuldt 2015). The St. Louis River runs through the reservation of the Fond du Lake Band of Lake Superior Chippewa, which is an active partner in restoration of the estuary.

Industrialization of the St. Louis River, which started in the 1800s, had a devastating impact on the health of the estuary. Over time, it became a dumping ground for untreated sewage and industrial waste (including wood waste from sawmills operating on the water's edge and toxic chemicals from paper mill discharges). Significant impacts on habitat occurred not only from these discharges, but also from logging operations, dredging of the navigational channel and filling of shallow wetlands. As a result, the St. Louis River was listed as one of 43 Great Lakes Areas of Concerns in 1987 by the International Joint Commission (IJC) of the United States and Canada.

Areas of Concern are defined in the Great Lakes Water Quality Agreement as "geographic areas … where significant impairment of beneficial uses has occurred as a result of human activities at the local level" (International Joint Commission 2012). Partners have been working diligently in recent years to complete actions necessary to delist the St. Louis River Area of Concern.

Today, the estuary is home to the Duluth-Superior Harbor, the nation's busiest freshwater harbor. Lake Superior draws many people to the region: more than a quarter of a million residents and 3.5 million tourists live, work and visit the area every year.

6.12.1 CLIMATE CHANGE IMPACTS WATER LEVELS IN THE ESTUARY

Currently, the most obvious and immediate impact of climate change in the St. Louis River Estuary and the Lake Superior shoreline is fluctuating water levels. Lake Superior influences water levels in the estuary on daily, seasonal and long-term time frames. Daily fluctuations occur due to storms and seiche and are influenced heavily by weather conditions. Storm surges or wind setups occur when winds blowing across the lake push the lake water in a certain direction. The pressure of the wind causes water levels to increase downwind. Water levels can change drastically under storm conditions. If atmospheric pressure changes dramatically or the wind stops suddenly during a wind setup, a seiche can occur. Seiche is a standing wave that oscillates back and forth in the lake basin. Seiche can also create significant changes in water levels as the wave moves back and forth. Though there is always upstream flow from the St. Louis River, the lower portion of the river is influenced by the Lake Superior seiche, and water flow sometimes changes direction. The estuary is defined as the portion of the river that is influenced by Lake Superior water levels.

Beginning in 2014, water levels have risen significantly in Lake Superior, and therefore in the estuary, as compared to the previous 15 years (Figure 6.12). Higher water levels significantly impact wetland habitats in the estuary, in many cases completely flooding out important habitat features. Higher water also impacts shorelines inhabited by creatures that rely on sandy beaches or mudflats, such as the Common Tern discussed below. In addition to generally higher waters, the estuary is also impacted by more frequent and intense storms in Minnesota, with higher wave action and storm surges battering sensitive habitats.

One incredibly damaging storm event happened in June 2012, when 5–10 inches of rain fell across northeast Minnesota and the Duluth area over two days. As a result, flows in the St. Louis River reached record rates and record crest heights (45,300 cubic feet per second and 16.6 feet, respectively) at the Fond du Lac dam in Scanlon, MN. For comparison, the mean monthly discharge in June for the period of record (1908–2020) is 3,620 cubic feet per second.

Natural systems are inherently capable of dealing with large episodic events through the natural process of flow and sedimentation. In the case of the 2012 event, major impacts were seen on the landside and floodplains surrounding the estuary, where human infrastructure was impacted. Changes in the river occurred as well, but they were not devastating effects. Perhaps the largest impact of more frequent and

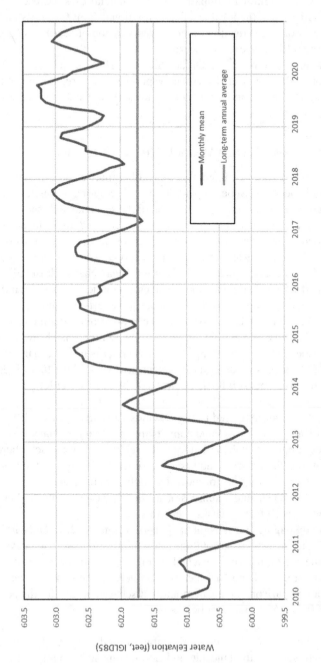

Data Source: U.S. Army Corps of Engineers Coordinated Monthly Mean Lakewide Average Water Levels from Lake Superior

FIGURE 6.12   Lake Superior water levels (Minnesota Land Trust).

intense storms in the St. Louis River Estuary is the uncertainty created: How will this habitat-limited environment respond? Will there be enough time for changes in location and type of habitats resulting from large events to adequately reestablish? These questions pose a significant challenge to natural resource managers as they plan conservation actions.

6.12.2   Minnesota Land Trust's Role in the Estuary Landscape

Over the past ten years, the Land Trust has supported the restoration and recovery of the St. Louis River in important ways. As a partner to MN DNR, the Land Trust used its restoration expertise to help develop its St. Louis River Restoration Initiative program and grow its project implementation experience by managing design and construction of large-scale restoration projects. Over time, agency staff developed the understanding and expertise necessary to implement large-scale projects themselves. The Land Trust's nimbleness, defined by its streamlined contracting procedures as compared to state agencies and its ability to work in both Minnesota and Wisconsin, allowed it to fill a niche in facilitating the bi-state and tribal stakeholders working in the estuary and getting projects done across state lines. Land Trust staff also serve as thought leaders, pushing the conservation agenda with partners further with unique ideas and solutions.

The Land Trust has managed completion of a number of large-scale restoration projects with partners in the estuary. Some of these projects involve species, including sturgeon and wild rice, that are significant culturally to the Ojibwe who have called the area surrounding the river home for hundreds of years.

Resource managers have been working since 1983 to restore the once abundant sturgeon population in the lower river. While adults are surviving, successful spawning is limited. Creation of hydropower dams blocked access to upstream spawning areas. At Chamber's Grove Park, the Land Trust managed a project to naturalize the river shoreline and create sturgeon spawning habitat. An existing sheet pile wall with an elevated boardwalk along the river's edge through the park limited access by people to the river for years. This area was heavily damaged in the 2012 flood, providing opportunity for partners to remove this blighted sea wall and accomplish multiple restoration goals in one location, including allowing for fluctuating water levels, increasing public access to the river and enhancing fish habitat. The restored park has become a local gem where people of all ages recreate and fish from the newly restored shoreline, and the location is now the second area in the lower river where sturgeon spawning habitat has been created.

Working in partnership with tribal resources managers at Fond du Lake Band of Lake Superior Chippewa and the 1854 Treaty Authority, as well as both Minnesota and Wisconsin Departments of Natural Resources, the Land Trust has also been helping to implement the wild rice restoration plan for the river. As described above, wild rice is sacred to the Ojibwe people and once grew abundantly in the estuary. The Land Trust has been managing efforts to seed prioritized locations throughout the estuary. This work is challenged by climate impacts, including production of wild rice in regional lakes that are used as seed sources for the river. Changing water levels have influenced the planting plans, with more seed being used throughout a wider range of depths to account for water level fluctuations.

Where agencies and municipalities can be limited by regulatory and political mandates, the Land Trust benefits from its position as a mission-driven organization. Its focus on restoring natural resources in the St. Louis River gives it the opportunity to choose what, where and how it wants to work. Seeing a need for visioning for future habitat conditions after the St. Louis River is delisted as an Area of Concern, the Land Trust developed an idea for establishing a framework for communication and collaboration in the estuary focused on natural resources management while considering intersections with community health and economic development. Together, these three pillars represent a sustainable landscape. The project, which was funded by the U.S. Fish and Wildlife Service, is bringing partners together to develop a forum for implementation of this framework and a future vision for the estuary that will allow us to address difficult topics in the estuary landscape, such as climate change.

6.13   INTERSTATE ISLAND – PROTECTING A VULNERABLE HABITAT AND SPECIES FROM FLOODING

Interstate Island, which straddles the Minnesota and Wisconsin state line in the lower St. Louis River, is home to one of only two Common Tern colonies in the Lake Superior watershed. Common Terns are a migratory species that overwinter in South America and return each May to nest on the island. They are listed as threatened in Minnesota and endangered in Wisconsin.

The island is designated as a Wildlife Management Area in Minnesota and a Wildlife Refuge in Wisconsin. It is also the only federally designated critical habitat for the Great Lakes Piping Plover, an endangered shorebird species. Minnesota and Wisconsin Departments of Natural Resources jointly manage the island.

The U.S. Army Corps of Engineers created Interstate Island in the 1930s out of dredge spoils from the Duluth-Superior Harbor. Because of extensive habitat loss from human use and development in areas where Common Terns historically nested in the area, it was decided in the late 1980s that Interstate Island would serve as a refuge for the Common Terns and the population was encouraged to move there to the island. Since that time, the tern colony has inhabited Interstate Island, along with a large colony (currently approximately 20,000 nesting pairs) of Ring-billed Gulls.

Habitat on the island is sandy and flat with shorelines scattered with driftwood that slope gently to the river's edge. Terns build their nests in sand and cobble in a fenced nesting area in the middle of the island, scratching out shallow depressions and surrounding them loosely with small rocks that may be present (Figure 6.13). Any shrubs and trees growing on the island can become perches for owls and other birds that might prey on tern chicks, so woody vegetation is controlled. Because of its low profile, water can dramatically change the amount of the island that is present above water.

In 2015, rising water levels in the estuary reached a point where the Common Tern nesting area was in dire threat of flooding. That year, the MN DNR found funds

FIGURE 6.13 Common Tern sitting on eggs in a nest on Interstate Island (Minnesota Land Trust).

FIGURE 6.14 Flooding at Interstate Island in July 2019 (Short, Elliott, Hendrickson, Inc.).

to complete an emergency project to elevate the nesting area. By 2019, however, even the elevated nesting area was threatened by high water levels with available habitat on the island reduced by half due to flooding (Figure 6.14).

The Land Trust became involved in the project in late 2018. Its role initially was connecting the dots between people and opportunities. Where MN DNR non-game wildlife managers had been limited by department funding and Wisconsin Department of Natural Resources was extremely staff limited, the Land Trust began submitting grant proposals and looking for opportunities within existing grants to get critical design work done. By April 2019, funds were in place to develop a site design and restoration of the island was on its way to becoming a required action for delisting of the Area of Concern. Shortly after, partners had secured US$1.4 million dollars to complete full restoration of the island with the Land Trust as project manager.

The restoration site team of bird experts and natural resource managers from both Minnesota and Wisconsin developed a set of habitat objectives to address further impacts from water levels with climate change: the island needed to be expanded to almost double its size with an increase in elevation such that 5.5 acres would be above water if historic high water levels were to occur, and the tern nesting area must sit another 3–4 feet higher than its current elevation. While the focus of the restoration was on the current population of Common Terns and Ring-billed Gulls occupying the island, the team also recognized that conditions on the island might change in the future with the changing climate. Shorebird habitat in the estuary and along Lake Superior is particularly vulnerable because of human use and its low-lying nature along shorelines. Accordingly, a focus was also given to providing suitable shorebird habitat at Interstate Island to other migrating species by providing more sandy shoreline with slopes suitable for shorebirds. Monitoring done as a component of the project validated this decision: in the first year, 16 species of shorebirds were identified using the island from spring migration through fall migration.

The concept design developed by the restoration site team included the desire to expand the island as much as possible into Wisconsin, given the constraints of a navigational channel bordering the island on the east and a large deep hole important for the river's fishery on the south. Funding had been secured from state and federal grants to restore flooded portions of the existing island footprint, to elevate the nesting area and to provide permanent fencing for the nesting area important for keeping gulls out. Opportunity to construct the largely expanded footprint came through a partnership with the U.S. Army Corps of Engineers. The Corps would construct the expansion using sand dredged from the navigation channel in the lower harbor through their routine operation and maintenance program – a win-win.

Construction of the restoration began in Spring 2019 with the import of gravel pit sand to fill the flooded areas of the island and regrade the existing shorelines (Figure 6.15). Work stopped in early May prior to the terns' return to the island to nest. Once all the terns left in August, the tern nesting area was elevated and fenced. The Corps's contractor then began construction of the island expansion area (Figure 6.16), which was completed in late November 2020. The few remaining construction elements were completed in Spring 2021. The completed restoration resulted in 8.4 acres above the ordinary high-water level (almost double the pre-restoration size), 6.7 acres above the design high-water level and an additional 840 feet of shoreline.

6.14 LESSONS LEARNED

Private land protection and restoration are important for climate action, by building resilient landscapes that benefit both wildlife habitat and human needs such as clean and abundant water and flood control. Conservation partnerships between public

FIGURE 6.15 Ring-billed Gulls present during construction in April 2019 (Minnesota Land Trust).

FIGURE 6.16 Construction of Interstate Island expansion area in October 2021 (Minnesota DNR).

agencies and NGOs targeting private lands have been proven a successful strategy to achieve landscape-level conservation goals. Within this constellation, Land Trusts are especially well positioned to serve as catalysts by helping to formulate a coherent vision and priorities on a landscape level and bringing together partners.

Because Land Trusts are citizen supported and have close connections to the local community, they often enjoy a level of trust that allows them to play the role as project champion and implementor. The Minnesota Land Trust experience demonstrates that land trusts can play a critical role in securing funding, connecting private landowners with tools and resources and providing the expertise and management capacity to implement projects on the ground.

To deliver on this work and expand Natural Climate Solutions, building relationships and trust matter. This takes time and sustained individual effort, whether it is the private landowner who protects and restores their property over decades or the resource manager who connects the right partners and completes projects throughout a career. Combined, these individual actions and web of relationships support the long game of landscape-level conservation and relentless incrementalism that is required to improve climate resiliency across the landscape.

Finally, supporting land conservation, whether through public or private funding or through conservation-friendly policies, produces concrete, tangible results for climate action and many other public benefits. Our lands and waters across the country hold the key to our food security, to our wildlife populations, to our health and quality of life and, in the end, to our ability to respond proactively to the increasing threats of climate change.

ACKNOWLEDGMENTS

Lead Editor: Ruurd Schoolderman, Program Manager Mississippi Headwaters

Reviewers: Kris Larson, Executive Director, Wayne Ostlie, Director of Land Protection

SPECIAL STUDY CONTRIBUTORS

Nick Bancks, Program Manager Blufflands *Planning for Biodiversity Resiliency – Heritage Brook Trout Streams Minnesota Blufflands*

Ruurd Schoolderman, Program Manager Mississippi Headwaters *Keeping Cold Water Cold – Mississippi Headwaters Cisco Refuge Lakes*

Beverly Rinke, Program Manager Wetlands and Grasslands, and Haley Golz, Private Lands Restoration Program Manager *Restoring Ecosystem Services – Keeping Water on the Land*

Virginia Breidenbach, Lake Superior Projects Coordinator *Coastal Habitat Protection in the St. Louis River Estuary*

LITERATURE CITED

Audubon (2021). St. Louis river important bird area. Retrieved from https://www.audubon.org/important-bird-areas/st-louis-river-estuary. Accessed June 2021.

Ahlering, M. et al. (2021). *Nature and climate solutions for Minnesota*. The Nature Conservancy.

Brice, J. S., Carrlson, K. M., Fields, S. P., Loesch C. R., Murano, R. J. D., Szymanski, M. L., Walker, J. A. (2017). 2017. *Prairie pothole joint venture implementation plan*. Denver, CO: U.S. Fish and Wildlife Service. 35 pages. https://ppjv.org/assets/pdf/PPJV_2017_ImplPlan_Sec2.pdf

Chismar, J. (2015, February 15). Climate change in Minnesota: 23 signs. *MPR News*. Retrieved from https://www.mprnews.org/story/2015/02/02/climate-change-primer. Accessed June 3 2021.

Cross, Timothy K., & Jacobson, Peter C. (2013). Landscape factors influencing lake phosphorus concentrations across Minnesota. *Lake and Reservoir Management*, 29(1), 1–12.

Hoxmeier, J. H., Dieterman, D. J., & Miller, L. M. (2015). Brook trout distribution, genetics, and population characteristics in the driftless area of Minnesota. *North American Journal of Fisheries Management*, 35(4), 632–648. doi: 10.1080/02755947.2015.1032451.

International Joint Commission (2012). Annex 1 of the Great Lakes water quality protocol of 2012. Retrieved from https://binational.net/wp-content/uploads/2014/05/1094_Canada-USA-GLWQA-_e.pdf. Accessed June 2021.

Jacobson, P. C., Fang, X., Stefan, H. G., & Pereira, D. L. (2013). Protecting Cisco (Coregonus artedi Lesueur) oxythermal habitat from climate change: Building resilience in deep lakes using a landscape approach. *Advances in Limnology*, 64, 323–332.

Jacobson, P. C., Stefan, H. G., & Pereira, D. L. (2010). Coldwater fish oxythermal habitat in Minnesota lakes: Influence of total phosphorus, July air temperature, and relative depth. *Canadian Journal of Fisheries and Aquatic Sciences*, 67(12), 2002–2013.

Minnesota Department of Natural Resources, Division of Ecological Services, and Minnesota Department of Natural Resources (2006). *Tomorrow's habitat for the wild and rare: An action plan for Minnesota wildlife, comprehensive wildlife conservation strategy*.

Minnesota Department of Natural Resources, Division of Ecological Services, and Minnesota Department of Natural Resources. *Watershed health assessment framework*. Retrieved from https://www.dnr.state.mn.us/whaf/index.html. Accessed June 28th, 2021.

Minnesota Department of Natural Resources (n.d.). *Historic mega-rain events in Minnesota*. Climate Summaries and Publications. Retrieved from https://www.dnr.state.mn.us/climate/summaries_and_publications/mega_rain_events.html. Accessed June 3, 2021.

Minnesota Department of Natural Resources (n.d.). Retrieved from https://gisdata.mn.gov/dataset/biota-cisco-refuge-lakes. Accessed June 9th 2021.

Minnesota Department of Natural Resources (2016). *Minnesota's wildlife action plan 2015–2025*. Division of Ecological and Water Resources, Minnesota Department of Natural Resources.

Minnesota Prairie Plan Working Group (2018). *Minnesota prairie conservation plan* (2nd edn.). Retrieved from https://www.nature.org/media/minnesota/mn-prairie-conservation-plan.pdf. Retrieved June 28th 2021.

MPR News (2021). *Minnesota Home Insurance Premiums Tripled as Extreme Weather Increased*. Retrieved from https://www.mprnews.org/episode/2021/03/25/minnesota-home-insurance-premiums-tripled-as-extreme-weather-increased. Accessed March 25, 2021.

Olmanson, L. (2021). HUC_08_Major_WS_Temperature_1985_2019 GIS Shapefile, data source. Retrieved from https://arcgis.dnr.state.mn.us/ewr/climatetrends.

Schuldt, N. (2015, October 6). Gitchi-Gami-ziibi: The Lake Superior river. *American Rivers*. Retrieved from https://www.americanrivers.org/2015/10/gitchi-gami-ziibi-the-lake-superior-river/. Accessed June 2021.

Summit Post (2021). *Public and private land percentages by state*. Retrieved from https://www.summitpost.org/public-and-private-land-percentages-by-us-states/186111. Accessed June 9th 2021.

The Conservation Fund (2020). Retrieved from https://www.conservationfund.org/impact/press-releases/2354-72-440-acres-of-working-forests-purchased-in-minnesota. Accessed June 11th 2021.

The Nature Conservancy (n.d.). *Natural climate solutions*. Retrieved from http://naturalclimatesolutions.org/

The Nature Conservancy (2016). Resilient and Connected Landscapes, Conservation Gateway. Retrieved from https://www.conservationgateway.org/ConservationByGeography/NorthAmerica/UnitedStates/edc/reportsdata/terrestrial/resilience/Pages/default.aspx. Accessed June 28th 2021.

The Nature Conservancy (n.d.). Resilient mapping tool. Retrieved from http://maps.tnc.org/resilientland/. accessed June 9th 2021.

Tipping, Robert G. et al. (2019, August). *Evaluation of temperature, streamflow, and hydrogeology impact on brook trout habitat*. University of Minnesota. Retrieved from https://www.lccmr.leg.mn/projects/2016/finals/2016_03k_summary_report.pdf.

United States Environmental Protection Agency (2016). *What climate change means for Minnesota*. EPA 430-F-16-025.

7 Sc̓iłpálqʷ
Biocultural Restoration of Whitebark Pine on the Flathead Reservation

Michael Durglo, Jr., Richard G. Everett, Tony Incashola, Jr., Maureen I. McCarthy, ShiNaasha H. Pete, Joshua M. Rosenau, Séliš-Q̓lispé Elders Cultural Advisory Council, Thompson Smith, Shirley Trahan, and Anne A. Carlson

CONTENTS

DOI: 10.1201/9781003048701-7

7.1 SČ̓IₗPÁLQᵂ: WHITEBARK PINE AND THE SÉLIŠ AND QₗISPÉ PEOPLE[1]

From the beginning of human time, the people of the Séliš (Salish or "Flathead") and Qₗispé (upper Kalispel or "Pend d'Oreille") nations have lived in the valleys and mountains of the Northern Rockies and adjoining areas (SQCC 2021). Across these vast Indigenous homelands, the ancient trails follow the ridgelines and crisscross the mountain passes. Tribal people have gone to those places for many reasons, including gathering foods and medicines, hunting, and making prayers and offerings. And the people have always known the tree that grows only there. In the Salish language, it is called sč̓iₗpálqᵂ. English speakers refer to it as whitebark pine. In scientific terminology, it is named *Pinus albicaulis.*

Whitebark pines can live for more than a thousand years. That's twice the five centuries since Columbus first landed in the Americas and initiated the transformation of the continent. But it is still a small fraction of the 12,000 years, or more, that the Séliš and Qₗispé have lived with sč̓iₗpálqᵂ. The tribes' collective memory holds a deep knowledge of many generations of whitebark pine, and how they have stood present, watching over the people from the ridgetops.

In this chapter, we describe the efforts of the Confederated Salish and Kootenai Tribes (CSKT) to save and restore sč̓iₗpálqᵂ, which is now threatened with extinction from the accelerating and intensifying effects of the climate crisis and disease. Too often, there is little communication and exchange between Indigenous communities working on issues of cultural survival and restoration, and scientists focused on their own research methodologies and approaches. On the Flathead Reservation, however, those disparate worlds are being brought together in a visionary effort to save whitebark pine.

Biocultural restoration on the Flathead Reservation goes beyond simply combining "culture and science" in a simplistic way. Building on this community's long history of educational projects that have bridged these disciplinary divides (CSKT-NRD 2021b, 2021a, 2021c), we are searching for and finding ways to ensure that it is the tribes' cultural relationship with sč̓iₗpálqᵂ that structures what we do and how we apply the powerful tools of Western science. This is not another case where the superficial trappings of Indigeneity are used as little more than a colorful logo on an otherwise conventional scientific methodology. Instead, tribal cultural ways are shaping every aspect of this endeavor, from the purpose and goals of the project to how we physically engage with the trees. We are taking care to ensure that scientific perspectives do not redefine our relationship with this tree, but rather that we are employing science as a tool to pursue our cultural objectives.

Before we describe our scientific work, let us offer a brief encapsulation of our cultural relationship with sč̓iₗpálqᵂ.

Perhaps because of the time-depth of our relationship, we have known sč̓iₗpálqᵂ in ways that are hard to fully explain to the public. It has to do with how we have lived and prayed, how in every aspect of our lives we have been deeply connected to the plants and animals around us. In our creation stories, the plants and animals make themselves into food and medicine. In giving of themselves, they made possible our coming into this world, our lives and our sustenance. We are forever indebted to them. At the heart of our cultures, reflected and reinforced in the prayers and ceremonies, human beings occupy a place of humility, relying upon the material and spiritual help of the plants and animals. In return, people are obligated to always treat plants and animals with respect. These relationships are full of joy and happiness in the gifts we receive each day.

The elders have said that it is from those creation stories, from the sense of gratitude to plants and animals, that the people must honor the plant or animal, the one who gave its life so that the people might live. So the ethic of avoiding waste stood at the center of the traditional way of life. As Qlispé elder Pete Beaverhead said in a 1975 recording, that the ethic held true for anything they gathered or killed:

"Kʷemt pentč u esyaʔ u es čteʔmim łu spʼiqałq u łu swewł u łu sťaťap łu skʷiskʷs łu stem. Esyaʔ u es ʔiłistm. Esyaʔ u es čteʔmim."

("They always used all of the berries, the fish, the things they killed, like the ruffed grouse or anything else they gathered or killed. They ate all of it. They used all of it.")

(P. Beaverhead, in Séliš-Qlispé Culture Committee archives (SQCC 2021))

In another 1975 recording, Séliš elder Louise Wisqn Vanderburg conveyed the same central principle, "Šeỷ łu sʔiʔiłis łu tsqsi … Esyaʔ še kʷeʔeys. Ta epł es xʷel. ("That was their food a long time ago … They took everything. They didn't throw anything away") (L.W. Vanderberg, in Séliš-Qlispé Culture Committee archives (SQCC 2021)).

The elders have told stories about sčiłpálqʷ. They remembered camps up in the mountains, near places like Čpáaqn, the mountain on the Reservation Divide. They told how the people would gather some of the heavy, dense cones and place them by the fire, where they roasted them until they opened up. The cones were full of seeds, called sčeyłp. And then the old people would warn the children that the seeds are rich and to not eat too many or they will get a belly ache! Séliš-Qlispé culture has always been characterized by close, careful, quiet observation of the natural world, and so our elders would say that only smxmxe, grizzly bears, can eat a lot of the oily nuts at one time, when they are fattening up for hibernation. And only grizzlies have jaws strong enough to crush open the dense cones.

The Séliš and Qlispé have also always known the bird called snałqʷ, the one known in English as Clark's nutcracker, and as *Nucifraga columbiana* in scientific terminology. This bird depends on sčiłpálqʷ—and sčiłpálqʷ also depends on snałqʷ. Among all the birds, only snałqʷ has a beak perfectly shaped to pry open the whitebark cones and extract the seeds. As scientists now also know, snałqʷ carries the seeds around and caches them here and there across many miles, retrieving many, but leaving some to germinate. The seeds left behind then grow, and in this way snałqʷ serves as the only way by which sčiłpálqʷ spreads across the mountains.

The elders tell us scientific methods and technologies can be good too, but only as long as we are careful to use the new ways in service to—under the guidance and direction of—the old ways of knowing, seeing and understanding. The elders have long encouraged us to master and employ the new scientific tools and approaches, but they have also warned that we must never lose the foundation of the traditional way of life: a spiritual and material relationship of respect with sčiɫpálqʷ, and all the other plants and animals, and the lands and waters.

7.2   INTRODUCTION TO THE CONFEDERATED SALISH AND KOOTENAI TRIBES AND THEIR LANDS

The CSKT include the Salish, Kootenai and Upper Kalispel Tribes, who originally lived and traded throughout the northwestern region of the lands now known as the United States into the southern part of the lands now known as Canada. Their home-lands include what are now portions of the states of Montana, Idaho and Oregon, and range into the Columbia River Basin in Washington State (CSKT-CRP 1996a). The Tribes ceded most of this land to the U.S. government on July 16, 1855, through the Treaty of Hell Gate, reserving the right to hunt, fish, graze and harvest in open and unclaimed lands within the Tribes' aboriginal territories (Devlin 2021; Bigart and Woodcock 1996); and also reserving from cession, as unceded sovereign land, certain areas including the Flathead Reservation in northwest Montana. The Reservation is located in the lower quarter of the Flathead River Basin, including the southern half of Flathead Lake and the Lower Flathead River Basin, including the southern half of Čɫq̓étkʷ (Flathead Lake) and Ntx̣ʷétkʷ (the Lower Flathead River). As of 2021, the Flathead Reservation is 1.317 million acres (532,971 ha) in size, of which 815,669 acres (330,090 ha) are owned by the Tribes as a whole, and 29,956 acres (12,123 ha) are owned in trust by Native American individuals (CSKT 2021).

As the first to organize a tribal government with a Constitution and a Corporate Charter under the Indian Reorganization Act of 1934, the CSKT are governed by a ten-person Tribal Council (CSKT 2021). In May 2020, there were 8,059 enrolled Tribal members with 5,360 living on the Reservation (CSKT-Annual Reports 2021). The total population of the Reservation was 31,394 in 2019 with American Indians accounting for approximately 32% of the entire Reservation population (U.S. Census Bureau 2012; CSKT-Annual Reports 2021).

The Tribes' mission is guided by traditional principles and values, which form the basis for all facets of tribal operations and services. Critical components include investing in their communities to ensure their ability to function as a completely self-sufficient society and economy, while simultaneously striving to provide sound environmental stewardship that preserves, perpetuates, protects and enhances nat-ural resources and ecosystems (CSKT-Annual Reports 2011). CSKT's vision is to maintain traditional principles and values in the ways they govern their reservation and represent themselves to the rest of the world; and in the ways they continue to preserve their right to determine their own destiny (CSKT-Annual Reports 2011). Today, the CSKT are recognized as a model of a self-sufficient sovereign Indigenous nation within the United States. Their tribal government offers a number of services to tribal members and is the chief employer on the Reservation (CSKT 2021).

7.3   CLIMATE IMPACTS TO MONTANA INCLUDING THE FLATHEAD RESERVATION

As the leaders of the world's nations have failed to effectively reduce the output of CO_2 and halt the emissions of other harmful industrial by-products, we Indigenous peoples find ourselves as leaders in the movement to prepare for a changing climate, and even more importantly, leading by following the ways and values of our ancestors, demonstrating an alternative path from the destructive ways of colonialism and capitalism. [CSKT Climate Change Strategic Plan, September 2013, updated April 2016 (CSKT-Climate 2016)]

CSKT has compiled detailed information of current and projected climate impacts to Montana and the Flathead Reservation, as part of their 2013 Climate Change Strategic Plan (updated in 2016) and as part of the new Climate Adaptation Plan scheduled for release in 2021 (CSKT-Climate 2016). Climate data and projected impacts to the Flathead Reservation are available through collaborating institutions. The Montana Climate Assessment (MCA), completed in 2017, assessed current and projected future climate change for all of Montana, including the northwestern region that encompasses the Flathead Reservation (MCA 2017; Whitlock et al. 2017). The Montana Climate Office (MCO) provides high-quality climate information and services to all Montanans (MCO 2021). According to the MCA, annual average temperatures across the state have risen between 1950 and 2015 by 2.7°F (1.5°C), with winter and spring temperatures rising by 3.0°F (1.7°C) over the same time period. On the Flathead Reservation, the total temperature increase has been 2.5°F (1.4°C), while the number of days above 90°F (32°C) per year has tripled in the last century (Whitlock et al. 2017).

Projections for average annual warming in Montana range from 4.5°F to 6.0°F (2.5–3.3°C) by mid-century (2050), depending on the greenhouse gas (GHG) emissions scenario. The lower range results from the moderating (RCP 4.5) scenario in which atmospheric CO_2 concentrations are stabilized by mid-century (through active mitigation measures) and the higher range is from the accelerating (RCP 8.5) scenario, which assumes emissions continue to rise on a "business as usual" path. By the end of the century, average annual temperatures in Montana are projected to increase by 5.6°F to 12.9°F (3.2–7.2° C), depending on the emissions scenario. These changes are expected to be consistent throughout the state, with the most significant temperature rises occurring in midsummer (Whitlock et al. 2017).

CSKT research partners from the Native Waters on Arid Lands (NWAL) program (McCarthy 2020) and the Montana Climate Office produced downscaled climate projections for the Flathead Reservation and extracted relevant data related to crop, livestock and forestry agriculture to inform the 2021 Climate Adaptation Plan (CSKT-Climate 2016; NWAL 2020; MCO 2021). These data are available for two elevations on the Flathead Reservation – the lower elevation (3,065 ft or 934 m) represents the Mission Valley where crops are grown, livestock are grazed and where the human population lives. The higher elevation (6,130 ft or 1,868 m) corresponds to a midpoint in the Mission Mountains, a range of the Northern Rockies located east of the valley, where snowpack accumulates and CSKT manages the forests of the Mission Wilderness. The highest peak in the Mission Range is 9,820 ft (2,994 m). Average annual historical temperatures for the period 1950–2000 in the Mission Valley were around 46°F (7.8°C); similarly, in Mission Mountains, average temperatures were around 32°F (0°C) at the midpoint. Average annual precipitation during this period was around 12 inches (30 cm) in the valley and approximately 60 inches (152 cm) in the mountains, with the latter falling primarily as snow.

Extracted climate projections show annual average temperatures to be 6–12°F (3–7°C) warmer than the historical average (depending on the emissions scenario) in both locations (NWAL 2020). Observations from the accelerating emissions scenario (the path the world is currently on) include the following: (1) average annual precipitation amounts are expected to stay near historic levels in the valley and increase around 5% in the mountains, with summer precipitation totals decreasing slightly and winter levels increasing, accordingly; (2) the frequency and intensity of extreme storms are expected to increase, especially in the mountains; (3) by 2100, the first snowfall will arrive about five weeks later than it has, historically; (4) significant warming temperatures in the mountains – with frost-free days doubling from ~75 to over 150 annually – will result in more rain and less snow during the winter, leading to significant declines in spring runoff; (5) growing degree days in the valley (defined as the first of at least six consecutive days with average temperatures over 50°F (10°C) will more than double by 2100; (6) the average number of hot days with temperatures at least 100°F (38°C) reaches over 30 in the valley by 2100, from a baseline close to zero, historically.

Overall climate change in Montana is leading to longer growing seasons with hotter temperatures and less spring runoff, which is expected to reduce crop yield for grass hay, alfalfa and barley and degrade the quality of rangeland forage for livestock (Whitlock et al. 2017; Zhao et al. 2017; Polley et al. 2017). Warmer stream temperatures may adversely impact fish populations and spawning rates (USGS 2021). Increased temperatures (especially warmer nighttime temperatures) stress native vegetation, allowing invasive plants to increase on rangelands, and allow crop and forest pests and pathogens to thrive (Finch 2012). One of the most impactful results of warming temperatures and reduced water supplies is increased risk of catastrophic wildfires. Warming temperatures have caused widespread changes in fire regime vegetation compositions, structures, functions and area extent, with increased wildland fire behavior (CSKT-Climate 2016; Barbero et al. 2015; Abatzoglou and Williams 2016; Coates et al. 2016). Wildfire risks and climate impacts are location dependent and occur at various scales, both temporally and spatially. Impacts in the high elevations in the Mission Mountains (home to the whitebark pine) differ significantly from those in the dry grass and sage steppe ecoregions of the Mission Valley, including areas where native vegetation has been replaced with introduced pasture grasses or planted crops. The combination of warming temperatures, reduced snowpack and changes in wildfire regimes is already impacting the health and survivability of the whitebark pine (Fryer 2002; Keane and Arno 1993; Keane et al. 2012).

7.4 CSKT WORLDVIEW AS A FOUNDATION FOR DEVELOPING TRIBAL RESPONSES TO CLIMATE CHANGE

The Séliš, Qĺispé and Ksanka people are the keepers of thousands of years of place-based traditional knowledge. Their relationship with the environment was, and continues to be, rooted in culture, tradition, experience and practice. Native stories and songs shared through generations provide evidence of significant ecological events that occurred as humans emerged onto a land already populated with plants, animals, rocks, soils and minerals. As in many other Indigenous communities, Traditional Ecological Knowledge (TEK) has guided climate adaptation among the Séliš, Qĺispé and Ksanka people for millennia and continues to underpin climate resilience today. The elders often remind us that it is now our responsibility to care for those who cannot care for themselves, including all of our human and non-human relations.

Within this expansive worldview, each of the three tribes of the Flathead Reservation, Séliš (Salish or "Flathead"), Qĺispé (upper Kalispel or "Pend d'Oreille") and Ksanka (Kootenai), is culturally unique and has its own belief system, while remaining similar in certain ways; fundamentally, each tribe holds traditional knowledge of the natural world, and each has profound respect for all of creation. Both traits have enabled the tribes to survive for thousands of years (CSKT-CRP 1996a), using subsistence patterns developed over generations of observation, experimentation and spiritual interaction with the natural world, which has in turn created a body of knowledge about the environment closely tied to seasons, locations and biology. This way of life is suffused with rich oral history and a spiritual tradition in which people respect the animals, plants and other elements of the natural environment. Through the teachings of elders (Figure 7.1), these tribal ways of life continue to this day (CSKT Cultural Preservation Office 2000).

Core tribal values are imbued with long-held cultural attitudes toward the land and its historical use of resources by ancestors and are based on the fundamental belief that humans and nature are one. They include (1) learning to live in harmony with each other and with the land; (2) understanding that our ancestors left this land to us and that we are only borrowing it from our children. For this reason, as we make decisions that affect the land, we must consider the consequences those decisions will have for the next seven generations, at least; (3) respecting the land, replacing what we take and taking only what we need – by respecting the land, we respect ourselves; (4) preserving spiritual and cultural practices and resources, acting on a spiritual basis when dealing with the environment, always being aware of and appreciating what we have and turning to the Creator when in need; (5) protecting Mother Earth and repaying the land for its gifts; (6) preserving an abundance of native animals, plants and fish; (7) maintaining traditional practices of gathering, hunting and fishing.

In deciding how best to respond to the existential threat of climate change, CSKT planning staff considered two questions that flowed directly from shared tribal values: "If you were given the opportunity to visit the Flathead Reservation 100 years

FIGURE 7.1   Tribal elders sharing knowledge and stories with CSKT children; from left to right: Bud Barnaby, Mike Durglo Sr. and Pat Pierre. *Photo credit*: Bernie Azure.

from now, what should the quality of life be like?" And, "What natural and human resources should the Reservation produce a century from now, and in what condition should the environment be?" The responses to these questions reflect cultural and traditional ties to the region extending back thousands of years.

To convey the integral nature of humanity and the natural and built worlds of the Salish, Kalispel and Kootenai peoples, the shared values of the three tribes have been formalized into a "Lifeways Wheel" (Figure 7.2). The wheel illustrates a tribal perspective for climate adaptation captured in the CSKT Climate Strategic Plan and the forthcoming CSKT Climate Adaptation Plan (CSKT-Climate 2016). Within the Lifeways Wheel, the circles of life and relationships radiate from the individual and family at the center outward to community and culture, tribe, nation and the world. The spokes on the Lifeways Wheel are the pillars of CSKT governance and represent tribal departments that have a key role in adapting to a changing climate: Forestry, Wildlife, Fish, Water, Infrastructure, People, Lands and Air. Within the Lifeways Wheel, all of the departments are woven together with braided sweetgrass, symbolizing healing, peace and tradition. The sector illustrations on the Lifeways Wheel highlight elements of tribal traditional life impacted by a warming world and changing climate (CSKT-Climate 2016), including forests populated by whitebark pine and huckleberries, wildlife relations, moose and bear, flowing streams home to bull trout and other fish, water flowing over the Seliš Ksanka Qĺispé Dam (formerly Kerr Dam), lifesaving infrastructure corridors over the main highway through the Reservation, our people (elders, youth and community members) honoring the river,

A. Individual / Family **B.** Community / Culture **C.** Tribe **D.** Nation **E.** World

FIGURE 7.2 The Confederated Salish and Kootenai Tribes Lifeways Wheel illustrating their worldview as a foundation for responding to climate change. *Illustration credit*: CSKT.

the fertile lands of Snyełmn Čłčewm (the Mission Valley) and the sweet air of our homelands.

As with all things, the people of the Confederated Salish and Kootenai Tribes follow the circle of life. The elders always say that all things are connected and that any impact or action on one resource will impact all. To better reflect this perspective, the CSKT Climate Change Advisory Committee (CCAC) analyzes impacts of climate disturbance on tribal resources as a whole. Over the last 50 years, the creation of tribal departments to manage natural and cultural resources has been shaped by Western (colonial settler) views of science, resource management and bureaucracy. Until recently, this influence has been an unavoidable part of tribal self-governance, but this is beginning to shift. This shift is being driven by the scope and extent of climate change impacts permeating and connecting all elements of life. Western reductionist approaches to managing people and the environment as separate sectors become ineffective in a living, breathing, interdependent system. Climate change affects all aspects of our shared environment at once. We are all in this together and we can only adapt, together. Climate change is increasingly seen as a problem of unintended consequences, which emerges when we fail to recognize that all things are connected.

The lesson of interconnection has been a part of traditional ecological knowledge for thousands of years. Climate change is not just an environmental problem resulting from increased atmospheric carbon dioxide in the atmosphere, but also a problem that extends from the denial that individual actions have a collective force and effect. These collective effects defy individual attempts to control and confine them, and so they demand from us our collective recognition and responsibility in order to manage them (Mitchell and Kwasset 2020). To that end, the CSKT CCAC is identifying climate resilience actions that transcend the department-by-department approach, drawing more from traditional knowledge and the idea of restoration and reciprocity in stewarding Mother Earth in a warming climate (Kimmerer 2011; Kimmerer 2013). CSKT climate adaptation approach is guided by the spirit of iɫaẃyeʔ (Eel-lah-wee-yeh), meaning great-great-grandparent, the whitebark pine tree who watched over the people and shared her wisdom for over a thousand years (Figure 7.3).

FIGURE 7.3 CSKT Climate Change Advisory Committee Chairman, Mike Durglo Jr., with iɫaẃyeʔ (Eel-lah-wee-yeh) or "great-great-grandparent"; an ancient whitebark pine that was estimated to be more than 1,000 years old when still alive. *Photo credit*: Mike Durglo Jr.

7.5 THE CSKT STRATEGIC CLIMATE PLAN AS A BEGINNING

In recognition of the existential threat posed by a warming climate, CSKT completed their first-ever Strategic Climate Plan for the Flathead Reservation in 2013 (CSKT-Climate 2016). The planning process employed a "Collective Impacts" model to better understand climate threats and galvanize adaptation efforts across tribal departments and programs. Through a series of meetings and workshops organized by the CCAC, tribal stakeholders, elders and participating subject matter experts identified management actions to address collective impacts and develop monitoring and evaluation metrics to assess progress toward creating the environment and quality of life envisioned for their children and their children 100 years in the future (CSKT-Climate 2016).

Tribal elders have recognized, prepared for and responded to changes in climate for millennia. It must be acknowledged that tribal adaptation to climate change as it has occurred over that period of time has been in response to natural variability, which was generally much slower and less extreme than what we are beginning to experience now, and what is predicted in coming decades. Nevertheless, the cultural knowledge and long experience of the Séliš, Qlispé and Ksanka provides a foundation for us as we move into an uncertain future. As such, the CSKT Climate Change Strategic and Adaptation Plans (CSKT-Climate 2016) draw upon that foundation in developing effective, culturally sensitive climate adaptation and mitigation strategies needed to ensure healthy human and ecological communities on the Flathead Reservation for generations to come. Within five years of completing the original climate strategy, tribal members recognized that climate change was unfolding far more quickly than originally projected by scientific models. In 2018, the Department of Historical Preservation – in collaboration with the Tribes' leadership, administration, elders, scientific leaders, community members, regional experts and other stakeholders – initiated a process to revise and update CSKT's Strategic Climate Plan and to identify ongoing and future adaptation actions. Historical information was adapted from the Flathead Reservation Comprehensive Resources Plan (CSKT-CRP 1996a, 1996b) and local climate change scenarios were adapted from Native Waters on Arid Lands (McCarthy 2020). Traditional Ecological Knowledge was provided by the Salish, Pend, d'Oreille Culture Committee, Kootenai Culture Committee and Historic Preservation/Cultural Preservation Department (CSKT-Climate 2016). Local impact assessments for several sectors (forestry, land, water, fish, air, water, wildlife, infrastructure, people and culture) were developed by CSKT Tribal Departments and partner organizations. Traditional knowledge, elder interviews, community survey results and supporting photographs and data underpin a circle of life approach to climate resilience (CSKT-Climate 2016). As with all other things, the new CSKT Climate Adaptation Plan is a "living document" designed to guide tribal resilience decisions and actions to conserve, restore and protect forests, land, water, air and all human and non-human inhabitants for generations to come. Restoring and protecting whitebark pine is an example of a climate adaptation action connecting the lifeways of forests, wildlife, water and people.

7.6   WHITEBARK PINE FORESTS AS AN EARLY
FOCUS FOR CSKT CLIMATE ADAPTATION

Beginning in 2013, the Flathead Reservation's ancient forests of sčiɫpálqʷ (whitebark pine) became an early focus for climate adaptation actions by CSKT. Whitebark pine is a gymnosperm, and a member of the division Coniferophyta (USDA-NRCS 2019). It is the only North American species of the "stone pine" group and can be a shade-tolerant, slow-growing, long-lived tree (Minore 1979). Mature whitebark pine trees will age to over 400 years, especially on harsh growing sites and older individuals in excess of 1,000 years have been found (Luckman et al. 1984; Perkins and Swetnam 1996). Whitebark pine is slow growing in both height and diameter and, although early to establish post-disturbance, will often be surrounded or topped out by faster growing species from the same establishment cohort (Arno and Hoff 1990).

Whitebark pine has the largest distribution of all five-needle white pines in North America (Tomback and Achuff 2010; Little and Critchfield 1969). It is found in the subalpine and treeline forests of the United States and Canada, including the northern Rocky Mountains, Great Basin, Sierra Nevada and Cascades and northern coastal ranges (Little and Critchfield 1969; Arno and Hoff 1990; McCaughey and Schmidt 2001). Whitebark pine forms extensive forests in the northern Rocky Mountains of the United States and it is also abundant on the eastern slope of the Cascades and Coast Ranges; mosaic-like stands tend to form at the northern end of its distribution (Arno and Hoff 1990). On the Flathead Reservation, whitebark pine is found generally on all slopes and aspects between 1,800 and 2,800 m elevation (5,900–9,200 ft); it will rarely establish either above or below this elevation band (Figure 7.4). Whitebark pine grows on poorly developed cold soils, and in deeper volcanic ash deposits (USDA 1975) and only starts cone production at 30–60 years of age. Cone production maximum is not attained until reaching a mid-upper canopy or open grown stature, around 125–250 years of age (Arno and Hoff 1990). The frequency of large, cone crops within a given stand varies regionally by climate and population level and ranges from yearly to every two or three to five years in many areas, with a few cones typically produced within a stand every year (Crone et al. 2011; Tomback et al. 2001; Arno and Hoff 1990).

Whitebark pine forests occur in two high-mountain biophysical settings, about which the Séliš, Q̓lispé and Ksanka people hold an intimate knowledge. On productive upper subalpine sites, whitebark pine is the major seral species that is replaced by the more shade-tolerant maninɫp (subalpine fir, *Abies lasiocarpa* [Hook.] Nutt.), t̓s̓ét̓p (Engelmann spruce, *Picea engelmannii* Parry ex. Engelm) or pɫtin̓éʔ (mountain hemlock, *Tsuga mertensiana* [Bong.] Carrière), depending on geographic region (Arno and Weaver 1990). These sites support upright, closed-canopy forests in the upper subalpine lower transition to timberline, just above or overlapping with the elevational limit of the shade-intolerant qʷqʷliʔt (lodgepole pine, *Pinus contorta* Douglas ex. Louden) (Arno and Weaver 1990; Pfister et al. 1977), allowing the two species to be codominant. Other forest codominants found with whitebark pine on these sites are cq̓eɫp (Douglas-fir, *Pseudotsuga menziesii* [Mirb.] Franco), kʷxʷtné (limber pine, *Pinus flexilis* James) and čtx̌ʷey cáqʷlš (alpine larch, *Larix lyalli*

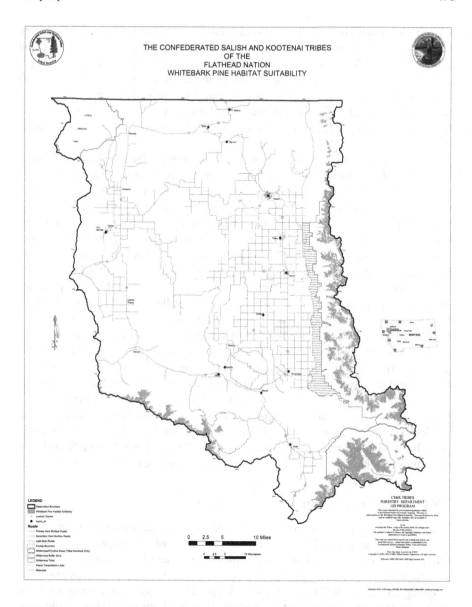

FIGURE 7.4 Map of Flathead Reservation depicting areas of suitable whitebark pine habitat. *Map credit*: Confederated Salish and Kootenai Tribes.

Parl) (Pfister et al. 1977). Fire regimes on these sites tend to vary from moderate-frequency/moderate-effects to low-frequency stand replacement events.

Whitebark pine can also successfully dominate high-elevation settings on harsh sites in the upper subalpine forests and at treeline on dry, high-elevation slopes (Arno 1986; Arno and Weaver 1990). Other species, such as subalpine fir, spruce and lodgepole pine, can occur on these sites as infrequent individuals with poor growth (Arno and Hoff 1990; Arno and Weaver 1990; Pfister et al. 1977). Alpine larch is often found on north-facing mature stand whitebark pine sites (Arno and Habeck 1972). Whitebark pine can also form krummholz and other typical high-elevation growth forms and stands in the alpine treeline ecotone (Arno and Hoff 1990; Tomback 1989) and as a minor component in lower subalpine sites (Cooper et al. 1991; Pfister et al. 1977). As tribal elders have long noted, with the open-grown nature of these stands, fires tend to be infrequent, low intensity and relatively small in size.

Whitebark pine forests form a wide variety of higher-elevation plant communities (Tomback and Kendall 2001). Most whitebark pine forests, especially the seral types, have relatively low diversity in vascular plants (Forcella 1978), with the majority of understory plant cover composed of ssipt (grouse whortleberry, *Vaccinium scoparium* Leiberg ex. Coville), stša (huckleberry, *Vaccinium membrenaceum* Douglas ex. Torr.), menziesia (*Menziesia ferruginea* Sm.), Hitchcock's smooth woodrush (*Luzula glabrata* [Hoppe ex. Rostk.] Desv. var. hitchcockii [Hämet-Ahti] Dorn) and slčéstye? (common beargrass, *Xerophyllum tenax* [Pursh.] Nutt), with sedge (*Carex rossii* Boott) and Geyer's sedge (*C. geyeri* Boott), Pink Mountain-heath (*Phyllodoce empertriformis* [Sm.] D. Don) and broadleaf arnica (*Arnica latifolia* Bong) – depending on geographical area, aspect and elevation – also form part of the understory vegetative component (Keane and Parsons 2010; Pfister et al. 1977). Other plants that may be occasional dominants include skʷlsé(łp) (kinnikinnick, *Arctostaphylos uva-ursi* [L.] Spreng.) Idaho fescue (*Festuca idahoensis* Elmer), Parry's rush (*Juncus parryi* Engelm.) and Wheeler bluegrass (*Poa nervosa* [Hook.] Vasey) (Arno and Weaver 1990; Aubry et al. 2008). High-elevation climax stands of whitebark pine support many unique alpine, subalpine and montane undergrowth species assemblages, some of which are only found in association with whitebark pine (Tomback and Kendall 2001). An early study (Forcella and Weaver 1977) found that whitebark pine forests had unexpectedly high biomass, but low productivity.

7.6.1 ECOLOGICAL INTERACTIONS AND IMPORTANCE

The whitebark pine is an important source of food for seed-eating birds and mammals (Keane et al. 2017). The high nutritional value in whitebark pine seeds is characterized by 21% protein (Lanner and Gilbert 1994) and fat content ranging from 28% (Robbins et al. 2006) to 52% (Lanner and Gilbert 1994). The seeds are large, weighing an average 0.175 g each, with thick seed coats (McCaughey and Tomback 2001; Tomback and Linhart 1990). This rich, tasty seed attracts sṅałqʷ (Clark's nutcracker, *Nucifraga columbiana*), a major seed disperser of the pine. Clark's nutcrackers each cache about 30,000–100,000 seeds a year in small, widely scattered caches, usually under 2–3 cm (0.75–1.25 in) of soil or gravelly substrate. Nutcrackers retrieve these seed caches during times of food scarcity and to feed their young. Cache sites selected by nutcrackers are often favorable for germination of seeds and survival of seedlings. Those caches not retrieved by the time the snow melts contribute to forest regeneration. Consequently, whitebark pine often grows in clumps of multiple trees, germinating from a single cache of two or more seeds (Tomback et al. 2001).

Up to 120 other species are known to utilize the whitebark pine (Keane et al. 2012). ʔiscč (pine squirrel or red squirrel, *Tamiasciurus hudsonicus*) cut down and store whitebark pine cones in their middens. Smx̣eyčn (Grizzly bear, Ursus arctos) and n̓łamqeʔ (American black bear, *Ursus americanus*) often raid squirrel middens for whitebark pine seeds, an important pre-hibernation food. Squirrels, kʷlkʷlé (northern flicker, *Colaptes auratus*) and n̓łqʷeyqʷay̓á (mountain bluebird, *Sialia currucoides*) often cavity nest in whitebark pines, and bear, tšeč or snečłčeʔ (elk, *Cervus canadensis*) and other vertebrates use mid-to-high-elevation whitebark pine communities as summer habitat.

7.6.2 Ecological Concerns

Whitebark pine forests are currently in decline throughout their range in North America because of the combined effects of historical and current fire exclusion management policies, mountain pine beetle (Dendroctonus ponderosae Hopkins) outbreaks since the 1980s and the invasive pathogen *Cronartium ribicola*, which causes the disease white pine blister rust in five-needle white pines (Keane and Arno 1993; Kendall and Keane 2001; Tomback and Achuff 2010). The loss of this high-elevation tree species poses serious consequences for upper subalpine ecosystems, both in terms of the impacts on biodiversity and in losses of both biotic and abiotic ecosystem processes and services (Tomback et al. 2001; Tomback and Achuff 2010).

7.6.3 White Pine Blister Rust

Most stands of whitebark pine across the species' entire natural range are infected with an introduced fungal pathogen, white pine blister rust (*Cronartium ribicola*). In the northern Rocky Mountains of the United States, whitebark pine mortality in some areas exceeds 90% (Keane et al. 2012). The blister rust also has overwhelmed commercially valuable western white pine in these areas and infected k̓ʷx̣ʷtné (limber pine, *Pinus flexilis*) populations as well (USDA-USFS 2021). There is currently no effective method for controlling the vectoring and disease effects of blister rust (USDA-USFS 2021; Keane et al. 2012). A single rust infection of a pine might establish and then spread for years. Once inside the pine needle, the fungus grows into the twigs and ultimately to the main stem of the tree. The damage is caused by the rust killing the cambium, causing a canker, ultimately girdling the stem which prevents water and nutrients from passing through the canker area; as a result, the twig, branch or stem dies. When the canker forms on the main stem, it causes top kill, often resulting in deformation or death of the tree. During the course of the disease, branch death and top kill significantly reduces cone production and tree vigor. A small number of trees (fewer than 5%) in most populations seem to possess genetic resistance to blister rust (Sniezko et al. 2018). Restoration efforts undertaken by forestry practitioners, tribal, federal government and private, in the northern Rocky Mountains involve harvesting cones from resistant whitebark pines, growing seedlings and planting seedlings in suitable sites.

Mountain pine beetle (Dendroctonus ponderosae) is a species of bark beetle native to western North America that may have also contributed to the widespread destruction of whitebark pine stands. Beetles lay their eggs and introduce pathogenic fungi into host trees, infecting many species of pine (Keane et al. 2012). This combination of larval development and fungal colonization is characteristically adequate to kill old or unhealthy trees. However, because of increased climate warming, the beetles have expanded their attacks to include younger, healthier trees (Kichas et al. 2020). Since the 1980s, the climate at high elevations has increased temperature enough for the beetles to reproduce within whitebark pine forest communities, often completing their life cycle within one year and enabling their populations to grow exponentially. Entire forest landscapes in the northern Rocky Mountains have become expanses of dead gray whitebark skeletons (Six and Bracewell 2015). In 2007, for example, the U.S. Fish and Wildlife Service estimated that beetles had killed whitebark pines across 500,000 acres (200,000 ha) in the west; while in 2009, beetles were estimated to have killed trees on 800,000 acres (320,000 ha), the most since record-keeping began (Tomback 2018).

7.6.4   FIRE SUPPRESSION

Incorporation of institutionally aggressive fire suppression has led to slow whitebark pine population declines during both the 20th and 21st centuries by interrupting the disturbance dynamics of stands without the occurrence of low- and moderate-intensity fire effects providing for forest resilience and health, and suppression of high-density stand insect-disease threats (Fryer 2002). In the absence of low-intensity periodic wildfire, whitebark pines in these stands are replaced by more shade-tolerant, fire-intolerant species such as maninɫp (subalpine fir, *Abies lasiocarpa*) and ťsťséɫp (Engelmann spruce, *Picea engelmannii*) (Fryer 2002; Keane and Arno 1993; Keane et al. 2012). This replacement often will lead to the development of a stand-replacement lethal fire regime, maximally perturbing whitebark forests. In addition, senescent and blister rust-infected pine trees are not destroyed by natural periodic ground fires, further diminishing the whitebark pine forest's vitality and survival (Keane et al. 2012; Kichas et al. 2021).

Rapid post-disturbance seed dispersal by Clark's nutcrackers combined with hardy seedling growth results in early whitebark pine community development after fire and other forest community disturbances (Tomback and Kendall 2001; Tomback et al. 2001). Furthermore, whitebark pine seedlings thrive in harsh, high-elevation droughty sites and may eventually act as early- to mid-seral nurse trees to more shade-tolerant conifers and vegetation (Tomback et al. 2001). At upper subalpine elevations, where it is common in many regions (Resler and Tomback 2008), whitebark pine helps regulate snow intercept and subsequent melt and reduce soil erosion (Farnes 1990). Whitebark pine is considered both a keystone species for encouraging community diversity and a foundation species for promoting ecological community stability (Ellison et al. 2005; Tomback et al. 2001; Tomback and Achuff 2010) and as an important ecosystem services provider. The loss of whitebark pine would potentially affect fire regimes, snowpack interceptions and local hydrology and other visual and recreational ecosystem services (Keane et al. 2012; Keane et al. 2017).

7.7 THE INDIGENIZED CONCEPT OF "BIOCULTURAL RESTORATION"

In considering how best to restore and sustain whitebark pine forests across the Flathead Reservation in the face of a shifting climate and other threats, the CSKT are adopting an Indigenized concept of restoration called "biocultural restoration." Biocultural restoration is the science and practice of restoring not only ecosystems but also human and cultural relationships to place, so that cultures are strengthened and revitalized along with the lands to which they are inextricably linked (Kimmerer et al. 2021; Greenlaw et al. 2009). CSKT has undertaken other projects based on this culturally informed approach, including the reestablishment of extirpated species (e.g., trumpeter swans) (Becker and Lichtenberg 2007), and a major restoration effort for the Jocko River (Explore the River) (CSKT Fish-Wildlife 2021). According to this concept, sčiłpálqʷ isn't just another tree. It is not a mute mass of wood and pulp, but a silent teacher of some of life's most valuable lessons that have been passed by the great-great-grandparent tree, iławiyeʔ, and many others in the forest (Figure 7.3).

Seedlings start their lives in high elevations where they are exposed to extreme variation in temperature, high winds and dense snow. The snowpack is so deep that sčiłpálqʷ evolved pliant branches able to bend under heavy snow without breaking. Sčiłpálqʷ spend several decades, 30–60 years, without ever bearing a single cone. After all that time spent growing in such a harsh environment, it might seem logical to presume that sčiłpálqʷ would be a very stingy organism, eking out a meager existence, and offering little to its fellow creatures. But sčiłpálqʷ proves otherwise. Full of rich fats and proteins, its pinion seeds provide food for dozens of animal species in the high-alpine forest. Of these animals, the tree has forged special relationships with sṅałqʷ, the Clark's nutcracker, and with ʔiscč, the pine squirrel. It might seem that these animals are greedy for taking so many of the pinions and hoarding them away underground; but this stockpiling behavior is balanced by the high number of caches that remain for other animals, like smx̣mx̣e, grizzly bears, to dig up and eat. This greediness is also balanced by those occasions when the caching of seed sprouts into a seedling so that sčiłpálqʷ can once again emerge.

Throughout time, the tribes have joined sṅałqʷ, and ʔiscč, to collect the seeds, and like the animals, their greediness is balanced. Natural resource workers collect the seeds and send them to the U.S. Forest Service Laboratory in Coeur d'Alene, Idaho, where disease-resistant individuals are identified. Collaborative teams from the U.S. Forest Service, Bureau of Land Management and CSKT later plant these healthy seedlings in high-elevation forests and burn scars from recent wildfires. The teams planted over 1,000 seedlings in 2019, and those that survive will outlive their planters by centuries.

These relationships demonstrate the ways that sčiłpálqʷ embodies humility, patience, perseverance, flexibility, generosity and reciprocity throughout its life cycle, and these lessons resonate in the culture of the Salish and Kootenai people. Despite these lessons, sčiłpálqʷ had not been a major priority of the tribes until climate planners began holding community gatherings, workshops and planning sessions deliberately focused on combining cultural practices and natural resource management efforts. It was only through community involvement and dialogue recognizing the interconnecting interests of human beings and the biotic community that sčiłpálqʷ reemerged as a species of tremendous importance. Sčiłpálqʷ has provided for the tribal community for many generations, and now the tribes are doing their part to preserve and protect sčiłpálqʷ, so that both lives are prolonged. The tribes are taking its lessons to heart.

This kind of dialogue is not new to Native people and is often very foreign and sometimes inaccessible for other communities. It is our hope that by sharing this story, others may use it as an example of how to build healthier relationships between human cultures and the biotic communities that support them. Sometimes, this seems like an impossible task, but as ancestors have reminded, the tribes in the past, and elders and sčiłpálqʷ continue to remind us today, "Never give up. Never give up."

7.8   THE PRACTICE OF RESTORING WHITEBARK PINE FORESTS AND ECOSYSTEMS ON THE FLATHEAD RESERVATION

Within the 1.3 million acres (526,091 ha) of the Flathead Reservation, the CSKT Forestry Department manages 256,757 acres (103,905 ha) of forest year-round (CSKT-Forestry 2000). The CSKT Forest Management Plan (2000) prioritizes new actions in forest management based on the principles of ecosystem management, all within the framework of traditional long-term forest management practices (CSKT-Forestry 2000). CSKT Forestry strives to manage sustainable forests for the future through the use of state-of-the-art computer models, satellite imagery and spatial analysis with geographic information systems (GIS). Within this focus, the Forest Management Plan states that "ecosystem management uses ecological, cultural, economic, social, and managerial principles to maintain and restore the ecological diversity and integrity of the reservation's surrounding forests" (CSKT-Forestry 2000).

The CSKT Forestry Department's goal is to restore Reservation forest habitat to be as similar to pre-European settlement as possible (CSKT-Forestry 2020). This goal considers the tribe's knowledge of different management practices of ecosystem restoration and how they apply to the different habitats. The 80-year goal from the Forest Management Plan is to enhance sustainable health, reestablish culturally traditional fire regimes and help manage and eliminate negative impacts from overstock, insect and disease on forest health (CSKT-Forestry 2000). Each goal seeks to address issues related to years of fire suppression and accelerating climate change, including increasing temperatures and decreasing annual precipitation (per above). Recognizing that balance between all-natural resources creates a sustainable forest, CSKT is committed to incorporating culture and traditional knowledge in all forest management decisions. CSKT Forestry not only seeks approval from the Tribal Council on proposed forest management directions and projects but also requests the guidance of the Séliš-Qlispé Culture Committee and the Kootenai Culture

Committee and from elders who voice their position on any project regarding the land. CSKT Forestry then selects the most relevant option to achieve ecosystem management goals (CSKT-Forestry 2020).

Climate change has had – and will continue to have – significant consequences for the health and restoration needs of whitebark pine forests throughout its distribution: both by acting as a stressor itself and by exacerbating other threats to this species (per above). Blister rust and mountain pine beetle outbreaks correlate with temperature increases at high elevations where mountain pine beetles have been historically absent; while a decrease in the occurrence of long, bitterly cold winters has created favorable conditions for mountain pine beetle breeding cycles (Bentz et al. 2009; Mahalovich et al. 2006). Over time, more frequent and severe drought events have weakened individual trees and made them more susceptible to disease, with beetles often present in the ecosystem now to finish off the job. Natural fire regimes have been disrupted significantly in the last century, disrupting natural processes that prevented the encroachment of tree species from ecosystems further downslope from competing with whitebark pine and contributing to the spread of more shade-tolerant species (White and CSKT 2010). Hotter temperatures and an increase in the duration of extremely hot days – both products of climate change – are the source of added ecological stress. Altogether, this combination of stressors has created conditions conducive to a stand-replacing, lethal fire outbreak in whitebark pine ecosystems. Consequently, CSKT is implementing key strategies in restoring whitebark pine to help the ecosystem rebalance by building resilience to each threat.

Whitebark pine and fire – both wildfire and the use of controlled burns as a long-term ecosystem management tool by the tribes – are two symbiotic natural resources that have strong cultural ties to the tribes and Forestry's focus and work. Traditionally, tribes used controlled burns to help sustain forest vegetation, and in so doing, helped maintain natural fire regimes across the landscape; with other ignition sources coming from lightning strikes and fires below the mountain running upslope (White and CSKT 2010). As a species, whitebark pine thrives after a fire has moved through an area and the previous forest canopy and competition are eliminated. Today, with whitepine blister rust and mountain pine beetle invasion, CSKT Forestry and CSKT Division of Fire are working together throughout the year to use controlled burns to clean out dead and woody downed material, to manage forest fuel loads and to prepare forest lands for restoration activities. Controlled burns help prepare an area in advance of planting and restoring forest stands through the use of understory or overcast burns; while simultaneously providing both the forest and forest workers a method for countering the effects from insects, disease and lethal fires in the whitebark pine ecosystem.

7.8.1 The Importance and Practice of Restoring Whitebark Pine Forests on the Flathead Reservation

Managing and restoring whitebark pine ecosystems is important to CSKT because of the benefits provided by this keystone species within whitebark pine forests and the lands below these forests. Not only are whitebark pine seeds a first food culturally, they are also important food sources for numerous wildlife species, including the grizzly bear. Ecologically, the loss of whitebark pine forests negatively impacts soil erosion and water quality for forest ecosystems and natural resources below (Keane et al. 2012).

As first step in the restoration process, Salish Kootenai College (SKC) initiated a student research project in 2013 to survey extant forest conditions, including tree age and fire frequencies, to characterize past forest conditions. SKC students also looked for evidence of human impacts to fire disturbance frequencies. Using this information, CSKT Forestry created their first maps and inventory of whitebark pine in 2016, in collaboration with researchers from the U.S. Forest Service, the Bureau of Land Management, and the Hi5 (whitebark pine) Working Group of the Crown Managers Partnership (CMP-Hi5 2019), a coalition of state, federal and provincial agencies working collaboratively with tribal and First Nations on shared natural resource priorities in the Crown of the Continent landscape (CMP 2021). These documents have guided CSKT Forestry in researching and mapping 105,000 acres (42,492 ha) of optimal subalpine habitat on the Reservation. Field observers found that there were signs that white pine blister rust was decimating whitebark pine throughout the region; but interestingly, our survey work has consistently documented trees in infected stands that are still healthy and resisting the fungus (Figure 7.5): the

FIGURE 7.5 A lone surviving whitebark pine tree in a stand of deceased whitebark pine trees on the Flathead Reservation. *Photo credit*: ShiNaasha Pete.

genetics of these resilient trees – known as "plus trees" (Keane et al. 2012) – are vital in the tribes' restoration work, and foster hope for future whitebark pine survival here and elsewhere.

Since 2016, CSKT reforestation specialists have continued to create an inventory of whitebark pine "plus trees." Each effort begins with the identification of locations that need to be scouted; using maps and existing inventories as the basis for on-the-ground searches for mature trees. In 2019, the Crown Managers Partnership (CMP 2021) began a partnership with a local geographic information system (GIS) firm, Map Monsters, to create detailed maps of the Reservation's whitebark pine stand conditions. The resulting conservation value map highlighted additional areas across the Reservation for expanded work on our inventory (CMP-Hi5 2019). The results of this landscape-scale collaborative effort are then laddering up to a national-scale restoration effort for this species, with the CSKT playing a leadership role at multiple spatial scales.

The next phase of long-term restoration work involves on-the-ground survey work in whitebark pine forest habitats, and, with a CSKT reforestation forester and Bureau of Land Management-certified tree climber(s), navigating rough logging roads by truck or ATVs in an attempt to get as close as possible to the survey areas. This can be very challenging given that most whitebark pine forests occur at high elevations that do not have (or, in the case of the CSKT Mission Wilderness areas, allow) roads. To travel in these areas requires packing in water, food, iPads, radios, tools, climbing gear and cages for young whitebark pine cones and hiking across survey areas searching for mature cone-bearing whitebark pines and potential "plus trees."

During the summer of 2019, SKC provided the CSKT-Forestry with three student interns to accompany a forester to scout whitebark pine habitat. Trekking over 20 miles of terrain in road-less areas during the day and camping at night, the team was able to identify and survey over 20 mature whitebark pine trees. When searching for a mature whitebark pine tree, teams look for trees that are cone-bearing and that appear safe to climb. Data collection then focuses on three different components: (1) basic information about the tree and its location, e.g., coordinates, date and time, diameter-at-breast-height, tree height, live crown ratio, whether it is a cone-bearing tree, which of the three growth forms this tree falls into and whether the tree is part of a whitebark pine clump; (2) disease information such as blister rust cankers, and any indication of flagging, swelling and oozing. If the tree has signs of a mountain pine beetle infestation, the stage of tree kill is recorded along with the estimated year of the last beetle attack recorded. Animal activity from grizzlies, red squirrels or Clark's nutcracker is marked in this area as well; (3) detailed stand information for the selected tree, including stand age, overstory composition of the three most common tree species, understory composition with the same three most common species and habitat type of the stand. If the tree shows no indications of stress, it becomes a "plus tree" and will continue to be monitored over time for growth and any changes indicating distress or disease in the stand.

The role of "plus trees" in the restoration of whitebark pine is essential. Currently, CSKT foresters collect cones from the very tops of each mature tree by caging a subset of the cones early summer before returning to collect them in the fall (Figure 7.6),

FIGURE 7.6   Caged whitebark pine cones after collection by CSKT tree climbers in autumn. *Photo credit*: ShiNaasha Pete.

per methods developed by the U.S. Forest Service, Rocky Mountain Research Station (Keane et al. 2017). Teams construct their own cages – which are designed to prevent seed depredation by squirrels – using duct tape and 0.25-inch (0.63 cm) fence mesh. The safety of CSKT Forestry-certified tree climbers in such remote areas is of paramount importance, with partners from the Bureau of Land Management or the U.S. Forest Service joining tribal staff at the end of each season to help collect cones. Each tree has an assigned number according to the year with a cone and cage count, with cones placed in burlap sacks, tagged with the tree number, date and time, and GPS coordinates.

Back at the CSKT Forestry Department, sacks of cones are stored at room temperature until cone collection for that season has been completed and the cones are dry and ready for processing. When cones are ready to have their seeds extracted, one sack at a time is processed. Cones – and all seeds – are processed by hand (Figure 7.7). The seeds from all cones in each sack (which originate from one tree only) are sealed in bags and marked with identifying information, before seed bags are weighed, recorded, sealed and stored (CSKT-Forestry 2000).

The germination process for whitebark pine seeds collected from "plus trees" comes next. In 2016, the first seeds collected on the Reservation were sent to the Coeur D'Alene Idaho U.S. Forest Service Nursery. Originating from nine

FIGURE 7.7 Whitebark pine seeds that have been removed from whitebark pine cones by CSKT restoration foresters. *Photo credit*: ShiNaasha Pete.

individual trees, these seeds were the first from CSKT lands to undergo testing for resiliency to the white pine blister rust fungus and complete their last cycle of inoculation testing this in 2021: during inoculation, whitebark pine seedlings are sprayed with white pine blister rust spores six to eight times over a two-year period to identify stock that are genetically resistant to the disease (Mahalovich et al. 2006). Since 2016, CSKT Forestry has successfully germinated their own whitebark pine seeds (Figure 7.8). Seeds are put through a strategic sterilization and scarifying process CSKT developed, which improves the rate of germination (CSKT-Forestry 2000). They are then sown in preparatory amendment soils and take two years for the seedlings to be ready to ship out for contracts with Crown Managers Partnership Whitebark Pine Working Group. The seedlings are stacked ten to a bundle with 50 bundles packed into a box and placed into the cooler until shipment. The seedlings will be distributed to Flathead National Forest, Glacier National Park, Bureau of Land Management, Lolo National Forest and here on the CSKT Reservation.

When planting whitebark pine, typically one hikes to the site after loading up planter bags that are either belted at the hips or slung over the back. When planting the seedlings, the selected locations would be the south-facing aspect of the mountain. To give the seedling some protection from natural elements, the seedlings

FIGURE 7.8 Confederated Salish and Kootenai Tribes' greenhouses where the Forestry Department germinates whitebark pine seedlings. *Photo credit*: ShiNaasha Pete 2020.

are planted on the north side of a rock or tree stump. CSKT Forestry had their first planting of whitebark pine in 2019 (Figure 7.9). The Liberty fire had burned over 400 acres in 2017 in the Jocko Primitive area and left perfect conditions on a 9-acre (3.6 ha) opening on a south-facing aspect, where past whitebark pine growth had been recorded previously. Over 1,900 seedlings were planted in June where the day started early and warm, ending with an early summer snow to cover the freshly planted whitebark pine seedlings. The CSKT planting crew departed in a muddy success, feeling accomplished and proud. Today, those seedlings are thriving with a 92% survival rate. CSKT, looking ahead, has 236 acres (106 ha) of whitebark pine restoration plots planned out in different areas in the Boulder Creek drainage on the east Flathead Lake shore mountain range. This area of the Reservation has a large inventory of "plus trees" and has been identified as a refugia area for this species.

Moving forward, CSKT Forestry's restoration goals for whitebark pine are to continue to scout, increase "plus tree" inventory, stock a whitebark pine seed bank for future fire, harvest restoration sites and create a high-elevation whitebark pine plantation; this includes applying proper management practices such as understory and broadcast burns, prepping stands by eliminating competition from non-whitebark pine species, daylighting activities to open the forest canopy and growing supplemental seedlings in CSKT Forestry green houses.

FIGURE 7.9 The first whitebark pine seedling planted in the mountains of the Flathead Reservation after the 2017 Liberty Fire. *Photo credit*: ShiNaasha Pete.

7.9 CULTURAL CONNECTIONS TO WHITEBARK PINE

Restoring whitebark pine forests across the Flathead Reservation has reconnected the people back to the meaning of this tree in tribal culture, rekindling traditional memories and stories like the ones shared in this chapter. The tribes now have significant opportunities to teach Native youth about the cultural importance of whitebark pine to carry into the future, through new internship and training programs for youth. One such program that has been recognized as exemplary throughout the state is Environmental Advocates for Global and Local Ecological Sustainability, or EAGLES. Awarded grants from the Dreaming Tree (a philanthropic foundation) have helped fund biological field technician internships for local high school and tribal college student internships, given the importance of teaching youth about ecosystem management and sustainable stewardship practices for the future. We are in need of saving our people, culture and land, and it is up to adults today to start teaching the young adults and youth (Figure 7.10). This is the most imperative goal, not only for forest management, but for whitebark pine as well.

Today, educational outreach is a necessity in the Environmental Science fields; but it is through the strength of CSKT's tribal culture that this keystone species will persevere through the hope, faith and effort that the restoration foresters, interns

FIGURE 7.10 CSKT schoolchildren learning about whitebark pine restoration efforts and climate change from members of the CSKT Forestry Department during a field trip to these ancient, high-elevation forests. *Photo credit*: Anne A. Carlson.

and students are putting into a species that they will never see thrive in their lifetimes. Nonetheless, the tribes feel strongly about sharing their efforts as a way of encouraging people to begin recognizing what is being lost today through global change. One way CSKT addressed this was by sharing our cultural connection to nature, and whitebark pine specifically, through a 2021 PBS documentary called "Ghost Forests" that featured work by CSKT on this species on the Flathead Indian Reservation (McCabe 2021). The film allowed viewers to accompany restoration foresters in the CSKT (among others) through the extraordinarily labor-intensive process of restoring whitebark pine forests across its distribution. Ongoing projects include a children's book and a virtual reality video across whitebark pine forests.

7.10 SEEING WITH BOTH EYES TO RESTORE AND PROTECT WHITEBARK PINE

"Sčiłpálqʷ: Biocultural Restoration by the Séliš (Salish) and Qlispé (upper Kalispel) Nations of the Flathead Reservation" illustrates the power of seeing with both eyes, or "two-eyed seeing," to conserve and restore whitebark pine. Two-eyed seeing is rooted in the concept that knowledge derived from Native wisdom and Western science, when woven together, allow the "seer" to create more holistic approaches for protecting our world for future generations (Marshall 2017; Colbourne, Rick et al. 2019; MacRitchie 2018; Peltier 2018). Underlying the two-eyed seeing approach are the concepts of humility and respect. No single person holds all knowledge – "humility" reminds us of that fact and "respect" allows different perspectives to be woven together.

The tribes' success story is not that they planted 2,000 trees, it is that they are restoring a lost connection, restoring understanding and restoring respect for the world that nurtures us. The story of the whitebark pine describes its longevity, decline, and now its revitalization mirrors the strength, loss and return of Indigenous ways of knowing and caring for Mother Earth. Indigenous people around the world have lived through major climate shifts before. They have survived, adapted and thrived due to their ability to see, learn and pass on wisdom about how to sustainably steward the forests, land, waters, air and wildlife upon which they depend. Biocultural restoration brings this traditional knowledge forward, and joins it with scientific observations, data and models of the ecological impacts of anthropogenic-driven global warming, to create a sustainable path that can reshape humanity's response to climate change. In turning to the teachings of Indigenous communities, teachings like the concept of reciprocity and respect for the environment, we can transition our path through climate change crisis from destruction to resilience (Mitchell and Kwasset 2020). Tribal communities' capacities for adaptation and restoration, which are rooted in millennia of observing and respecting the world around us, offer the world hope and wisdom.

Fundamental to this path are the relationships of people to place and the concept of reciprocity. In the words of Robin Wall Kimmerer of the Citizen Potawatomi Nation,

> For much of human's time on the planet, before the great delusion, we lived in cultures that understood the covenant of reciprocity, that for the Earth to stay in balance, for the gifts to continue to flow, we must give back in equal measure for what we take.
>
> **(Kimmerer 2011)**

Reciprocity requires we focus on the needs of the natural world as much as we do on our human needs. Every gift received from the natural world requires we give back in equal or greater measure. Restoration becomes an act of love for the world that sustains us and our gift to the seven generations that follow in our footsteps.

ACKNOWLEDGMENTS

The authors would like to express their deep gratitude to: the Tribal Council of the Confederated Salish and Kootenai Tribes; the Bureau of Indian Affairs Tribal Resilience Program for multiple funding awards from 2018 through 2021; the Roundtable of the Crown of the Continent Adaptive Management Initiative through the Kresge Foundation; the Great Northern Landscape Conservation Cooperative; Dreaming Tree; the Wilderness Society; the Crown Managers Partnership (CMP) and members of the CMP Hi5 Working Groups for additional funding and/or partnership; and all members of the CSKT Climate Change Advisory Committee. The CCAC is grateful to Ron Oden (Ron Oden Designs) for his artistic rendering to the Lifeways Wheel. Local climate change scenarios were provided by the Native Waters on Arid Lands Program (USDA/NIFA AFRI 2015-69007-23190) and the Montana Climate Office. The contributors would like to express their special appreciation for the work of our colleague Shirley Trahan, senior translator/transcriber and language advisor for the Séliš-Q̓lispé Culture Committee. Shirley worked through Salish-language recordings of tribal elders from the 1970s and provided our project with something of the greatest importance: bilingual transcripts of passages relating to sčiɫpálqʷ (whitebark pine) and its importance in Séliš-Q̓lispé culture. *Lemlmtš* (thank you) Shirley. Most of all, we are indebted to the Séliš-Q̓lispé Culture Committee and Elders Advisory Council. Since the CSKT's reestablishment of the Elders group as an organized entity in 1975, the elders have generously shared the stories, the cultural knowledge and information, the wisdom and perspective that has been handed down from the x̣ʷlčmússn, the ancestors, from the beginning of human time. This chapter is based on that foundation.

NOTE

1. Information in the chapter relating to the Salish Language and to the culture and history of the Sélis (Salish or "Flathead") and Q̓lispé (Upper Kalispel or "Pend d'Oreille") people is courtesy of the Séliš-Q̓lispé Culture Committee, a Department of the Confederated Salish and Kootenai Tribes.

LITERATURE CITED

Abatzoglou, John T., & Williams, A. Park (2016). Impact of anthropogenic climate change on wildfire across Western US forests. *Proceedings of the National Academy of Sciences*, *113*(42), 11770–11775. doi: 10.1073/pnas.1607171113. Checked on 11 Aug 2021.

Arno, S. F. (1986). Whitebark pine cone crops—A diminishing source of wildlife food? *Western Journal of Applied Forestry*, *1*(3), 92–94.

Arno, S. F., & Habeck, J. R. (1972). Ecology of Alpine larch [Larix Lyallii Parl]. In the Pacific Northwest. *Ecological Monographs*, 42: 417–450.

Arno, S. F., & Hoff, R. J. (1990). Pinus Albicaulis Engelm. Whitebark pine. In *Silvics of North America*. Washington DC: Department of Agriculture, Forest Service, *1*, Conifers:268–79. Agriculture Handbook 654 654.

Arno, S. F., & Weaver, T. (1990). Whitebark pine community types and their patterns on the landscape. In *Proceedings of the Symposium on Whitebark Pine Ecosystems: Ecology and Management of a High-Mountain Resource*. General Technical Report INT-270 (pp. 97–105). Ogden, UT: USDA Forest Service, Intermountain Research Station.

Aubry, C., Goheen, D., Shoal, R., Ohlson, T., Lorenz, T., Bower, A., ... Sniezko, R. A. (2008). *Whitebark pine resto-ration strategy for the pacific northwest 2009–2013*. Region 6 Report. Portland, OR: USDA Forest Service, Pacific Northwest Region.

Barbero, R., Abatzoglou, J. T., Larkin, N. K., Kolden, C. A., & Stocks, B. (2015). Climate change presents increased potential for very large fires in the contiguous United States. *International Journal of Wildland Fire*, *24*(7), 892. doi: 10.1071/WF15083. Checked on 11 Aug 2021.

Becker, D. M., & Lichtenberg, J. S. (2007). An update on the trumpeter swan reintroduction on the flathead reservation. *Tribal wildlife Management Program*. Pablo, MT: Confederated Salish & Kootenai Tribes. Retrieved from https://www.fws.gov/upload-edFiles/TrumpeterSwanPaper.pdf. Checked On 11 Aug 2021.

Bentz, Barbara, Logan, Jesse, MacMahaon, Jim, Allen, Craig, Ayres, Matt, Berg, Allan Carroll et al. (2009). *Bark beetle outbreaks in Western North America: Causes and consequences*. Salt Lake City, UT: University of Utah Press.

Bigart, Robert, & Woodcock, Clarence (Eds.) (1996). *In the Name of the Salish & Kootenai Nation: The 1855 Hell Gate Treaty and the Origin of the Flathead Indian Reservation.* Original printing 1996, 2nd printing 2008. Pablo, MT: Salish Kootenai College Press.

CMP (2021). Crown managers partnership. Retrieved from https://www.crownmanagers.org/. Checked on 11 Aug 2021.

CMP-Hi5 (2019). Crown of the continent high five working group technical team. Crown of the continent ecosystem whitebark pine restoration strategy: Pilot project summary. *Crown Managers Partnership.* Retrieved from https://www.crownmanagers.org/five-needle-pine-working-group. Checked on 11 Aug 2021.

Coates, Peter S., Ricca, Mark A., Prochazka, Brian G., Brooks, Matthew L., Doherty, Kevin E., Kroger, Travis, ... Casazza, Michael L. (2016). Wildfire, climate, and invasive grass interactions negatively impact an Indicator species by reshaping sagebrush ecosystems. *Proceedings of the National Academy of Sciences, 113*(45), 12745–12750. doi: 10.1073/pnas.1606898113. Checked 11 Aug 2021.

Colbourne, Rick, Moroz, Peter, Hall, Craig, Lendsay, Kelly, & Anderson Robert, B. (2019). Indigenous works and two eyed seeing: Mapping the case for indigenous-led research. *Qualitative Research in Organizations and Management: An International Journal, 15*(1), 68–86. doi: 10.1108/QROM-04-2019-1754. Checked on 11 Aug 2021.

Cooper, S. J., Neiman, K. E., & Roberts, D. W. (1991). *Forest Habitat types of northern Idaho: A second approximation.* General Technical Report INT-236. Ogden, UT: USDA Forest Service, Intermountain Research Station.

Crone, Elizabeth E., McIntire, Eliot J. B., & Brodie, Jedediah (2011). What defines mast seeding? Spatio-temporal patterns of cone production by whitebark pine: Spatio-temporal patterns of whitebark pine mast. *Journal of Ecology*, 99, 438–444. doi: 10.1111/j.1365-2745.2010.01790.x. Checked 11 Aug 2021.

CSKT (2021). Confederated Salish & Kootenai tribes of the flathead reservation. Retrieved from http://www.csktribes.org. Checked 11 Aug 2021.

CSKT climate change strategic plan (2013). Pablo, MT: Confederated Salish and Kootenai Tribes of the Flathead Reservation. Retrieved from http://www.csktribes.org/CSKTClimatePlan.pdf. Checked on 11 Aug 2021.

CSKT Cultural Preservation Office (2000). Cultural resource overview for the U.S. Fish and Wildlife Service western Montana management properties. Retrieved from https://csktribes.org/history-and-culture/cultural-preservation. Checked 11 Aug 2021.

CSKT Fish-Wildlife (2021). Explore the river: Bull trout, tribal people, and the Jocko River. *CSKT Fish, Wildlife, Recreation & Conservation.* Retrieved from http://fwrconline.csktnrd.org/Explore/ExploreTheRiver/. Checked 11 Aug 2021.

CSKT-Annual Reports (2011). Confederated Salish & Kootenai tribes' annual report 2011. *Confederated Salish and Kootenai Tribes.* Retrieved from https://csktribes.org/government/annual-reports. Checked 11 Aug 2021.

——— (2021). CSKT annual report 2019–2020. Retrieved from https://csktribes.org/component/rsfiles/files?folder=CSKT%2BAnnual%2BReports. Checked 11 Aug 2021.

CSKT-Climate (2016). *The confederated Salish and Kootenai tribes of the flathead reservation: Climate change strategic plan (Sep 2013, updated Apr 2016).* Pablo, MT: Confederated Salish and Kootenai Tribes. Retrieved from http://csktclimate.org. Checked 11 Aug 2021.

CSKT-CRP (1996a). Flathead reservation comprehensive resources plan volume 1 existing conditions. Retrieved from http://csktclimate.org/downloads/Comp_Plan/Comp%20Plan%20Volume%201.pdf. Checked 11 Aug 2021.

——— (1996b). Flathead reservation comprehensive resources plan volume II policies. Retrieved from http://csktclimate.org/downloads/Comp_Plan/Comp%20Plan%20Volume%202.pdf. Checked 11 Aug 2021.

CSKT-Forestry (2000). Flathead Indian reservation Forest management plan: An ecosystem approach to tribal Forest management. Retrieved from http://www.csktribes.org/component/rsfiles/download?path=Forestry%252Ffmp05.pdf. Checked 11 Aug 2021.

————— (2020). Tribal forestry. Retrieved from http://www.csktribes.org/natural-resources/tribal-forestry. Checked 11 Aug 2021.

CSKT-NRD (2021a). Explore the river – Bull trout, tribal people, and the jocko river. *CSKT's Online Educational Resources.* Retrieved from http://fwrconline.csktnrd.org/Explore/index.html. Checked 11 Aug 2021.

————— (2021b). Fire on the land: Native people and fire in the northern Rockies. *CSKT Online Educational Resources.* Retrieved from http://fwrconline.csktnrd.org/Fire/index.html. Checked 11 Aug 2021.

————— (2021c). Living landscapes – Culture, climate science & education in tribal and native communities. *CSKT's Online Educational Resources.* Retrieved from https://www.skclivinglandscapes.org. Checked 11 Aug 2021.

Devlin, Sarah (2021). The Salish tribe – The Treaty of Hellgate. *Salish Tribe History.* Retrieved from https://salishtribe.wordpress.com/location-history/treaty-of-hellgate/. Checked 11 Aug 2021.

Ellison, A. M., Bank, M. S., Clinton, B. D., Colburn, E. A., Elliott, K., Ford, C. R., ... Webster, J. R. (2005). Loss of foundation species: Consequences for the structure and dynamics of forested ecosystems. *Frontiers in Ecology and the Environment, 3*(9), 479–486.

Farnes, P. E. (1990). SNOWTEL and snow course data: Describing the hydrology of whitebark pine ecosystems. In *Proceedings of the— Symposium on Whitebark Pine Ecosystems: Ecology and Management of a High-Mountain Resource.* General Technical Report INT-270 (pp. 302–305). Ogden, UT: USDA Forest Service, Intermountain Research Station.

Finch, Deborah M. (2012). Climate change in grasslands, shrublands, and deserts of the Interior American West: A review and needs assessment. Gen. Tech. Rep. RMRS-GTR-285. Fort Collins, CO: United States Department of Agriculture, Forest Service, Rocky Mountain Research Station. Retrieved from https://www.fs.usda.gov/rmrs/publications/climate-change-grasslands-shrublands-and-deserts-interior-american-west-review-and. Checked 11 Aug 2021.

Forcella, F. (1978). Flora and chorology of the Pinus Albicaulis- Vaccinium scoparium association. *Madrono, 25,* 139–150.

Forcella, F., & Weaver, T. (1977). Biomass and productivity of the subalpine Pinus Albicaulis: Vaccinium scoparium association in Montana, USA. *Vegetation,* 95–105.

Fryer, J. L. (2002). Pinus albicaulis. *Fire Effects Information System (FEIS).* Retrieved from https://www.fs.fed.us/database/feis/plants/tree/pinalb/all.html. Checked 11 Aug 2021.

Greenlaw, Suzanne, Knowlden, Smantha, Landis, Catherine, & Talucci, Tyler (2009). Biocultural restoration in an urban watershed. *Onandaga Creek.* Syracuse, NY: Onandaga Environmental Institute. Retrieved from https://oei2.org/wp-content/uploads/2018/12/biocultural_restoration_project.pdf. Checked 11 Aug 2021.

Keane, R. E., & Arno, S. F. (1993). Rapid decline of whitebark pine in western Montana: Evidence from 20-year remeasurements. *Western Journal of Applied Forestry, 8*(2), 44–47.

Keane, R. E., Holsinger, L. M., Mahalovich, M. F., & Tomback, D. F. (2017). Restoring whitebark pine ecosystems in the face of climate change. *Gen. Tech. Rep.* Fort Collins, CO: United States Department of Agriculture, Forest Service, Rocky Mountain Research Station. Retrieved from https://www.fs.fed.us/rm/pubs_series/rmrs/gtr/rmrs_gtr361.pdf. Checked 11 Aug 2021.

Keane, R. E., & Parsons, R. (2010). Restoring whitebark pine forests of the northern Rocky Mountains. *Ecological Restoration, 28*(1), 56–70.

Keane, R. E., Tomback, D. F., Aubry, C. A., Bower, A. D., Campbell, E. M., Cripps, C. L. et al. (2012). A range-wide restoration strategy for whitebark pine (Pinus Albicaulis). *Gen. Tech. Rep. RMRS-GTR-279*. Fort Collins, CO: United States Department of Agriculture, Forest Service, Rocky Mountain Research Station. Retrieved from http://www.fs.fed.us/rm/publications. Checked 11 Aug 2021.

Kendall, K. C., & Keane, R. E. (2001). Whitebark pine decline: Infection, mortality, and population trends. In W. Pine (Ed.), *Communities: Ecology and restoration* (pp. 221–242). Washington DC: Island Press.

Kichas, Nickolas E., Hood, Sharon M., Pederson, Gregory T., Everett, Richard G., & McWethy, David B. (2020). Whitebark pine (Pinus Albicaulis) growth and defense in response to Mountain Pine beetle outbreaks. *Forest Ecology and Management*, *457*(February), 117736. doi: 10.1016/j.foreco.2019.117736. Checked 11 Aug 2021.

Kichas, Nickolas E., Trowbridge, Amy M., Raffa, Kenneth F., Malone, Shealyn C., Hood, Sharon M., Everett, Richard G., … Pederson, Gregory T. (2021). Growth and defense characteristics of whitebark pine (Pinus Albicaulis) and lodgepole pine (Pinus contorta Var Latifolia) in a high-elevation, disturbance-prone mixed-conifer Forest in northwestern Montana, USA. *Forest Ecology and Management*, *493*(August), 119286. doi: 10.1016/j.foreco.2021.119286. Checked 11 Aug 2021:.

Kimmerer, Robin (2011). Restoration and reciprocity: The contributions of traditional ecological knowledge. In D. Egan, E. E. Hjerpe & J. Abrams (Eds.), *Human dimensions of ecological restoration* (pp. 257–276). Washington, DC: Island Press/Center for Resource Economics. doi: 10.5822/978-1-61091-039-2_18. Checked 11 Aug 2021.

Kimmerer, Robin Wall (2013). Returning the gift. *Center for Humans & Nature*: Expanding Our Natural & Civic Imagination (blog). Retrieved from https://www.humansandnature.org/earth-ethic-robin-kimmerer. Checked 11 Aug 2021.

Kimmerer, Robin, Wall, Brittani Orona, & Reed, Charley (2021). Practicing biocultural restoration. Presented at the 2021 Biocultural Restoration Event, Virtual. Retrieved from https://www.rrnw.org/biocultural/. Checked 11 Aug 2021.

Lanner, R. M., & Gilbert, B. K. (1994). Nutritive value of White- Bark Pine seeds, and the question of their variable dormancy. In *Proceedings of the—international Workshop on Subalpine Stone Pines and Their Environment: The State of Our Knowledge* (pp. 206–211). General Technical Report INT-GTR-309. Ogden, UT: USDA Forest Service, Intermountain Research Station.

Little, E. L. Jr., & Critchfield, W. B. (1969). *Subdivisions of the Genus Pinus*. USDA Forest Service miscellaneous Publication 1144. Washington DC: United States Department of Agriculture, Forest Service.

Luckman, B. H., Jozsa, L. A., & Murphy, P. J. (1984). Living seven-hundred-year-old Picea engelmannii and Pinus Albicaulis in the Canadian Rockies. *Arctic and Alpine Research*, *16*(4), 419–422. doi: 10.2307/1550903. Checked 11 Aug 2021.

MacRitchie, Sarah (2018). Bridging western and indigenous knowledges: Two-eyed seeing and the development of a country food strategy in the Northwest Territories. *Masters, Graduate School of Public and International Affairs, Ottawa, Canada: University of Ottawa*. Retrieved from https://ruor.uottawa.ca/bitstream/10393/37576/4/MACRITCHIE,Sarah.20181.pdf. Checked 11 Aug 2021.

Mahalovich, Mary F., Burr, Karen E., & Foushee, David L. (2006). Whitebark pine germination, rust resistance, and cold hardiness among seed sources in the inland northwest: Planting strategies for restoration. In *National Proceedings: Forest and Conservation Nursery Associations – 2005* (pp. 92–101). Proceedings of the RMRS-P-43. Fort Collins, CO: United States Department of Agriculture, Forest Service, Rocky Mountain Research Station.

Marshall, Albert (2017). Two-eyed seeing – Elder albert Marshall's guiding principle for inter-cultural collaboration Presented at the Climate Change. In *Presented at the Climate Change, Drawdown & the Human Prospect: A Retreat for Empowering our Climate Future for Rural Communities*, Pugwash, Nova Scotia, Canada, October 28. Retrieved from http://www.integrativescience.ca/uploads/files/Two-Eyed%20Seeing-AMarshall-Thinkers%20Lodge2017(1).pdf. Checked 11 Aug 2021.

MCA (2017). Montana climate assessment. *2017 Montana Climate Assessment*. Retrieved from https://montanaclimate.org. Checked 11 Aug 2021.

McCabe, B. (2021). *Ghost forests*. PBS. Retrieved from https://www.montanapbs.org/programs/ghost-forests/. Checked 11 Aug.

McCarthy, Maureen (2020). Native waters on arid lands. *NWAL*. Retrieved from https://nativewaters-aridlands.com. Checked 11 Aug 2021.

McCaughey, Ward W., & Schmidt, Wyman C. (2001). Taxonomy, distribution, and history. In D. F. Tomback, S. F. Arno & R. E. Keane (Eds.), *Whitebark pine communities: Ecology and restoration* (pp. 29–40). Washington DC: Island Press.

McCaughey, W. W., & Tomback, D. F. (2001). The natural regeneration process. In W. Pine (Ed.), *Communities: Ecology and restoration* (pp. 105–120). Washington DC: Island Press.

MCO (2021). Montana Climate Office *University of Montana, Forest, Montana, & Conservation Experiment Station*. Retrieved from https://climate.umt.edu. Checked 11 Aug 2021.

Minore, D. (1979). *Comparative autecological characteristics of northwestern tree species: A literature review*. General Technical Report PNW-87. Portland. OR: USDA Forest Service, Pacific Northwest Forest and Range Experiment Station.

Mitchell, S., & Kwasset, W. H. (2020). Indigenous prophecy and Mother Earth. In A. E. Johnson & K. K. Wilkinson (Eds.), *All we can save: Truth, courage, and solutions for the climate crisis*. New York, NY: One World.

NWAL (2020). Native waters on arid lands climate projections. Retrieved from https://native-waters-aridlands.com/climate-projections/. Checked 11 Aug 2021.

Peltier, Cindy (2018). An application of two-eyed seeing: Indigenous research methods with participatory action research. *International Journal of Qualitative Methods, 17*(1): 160940691881234. doi: 10.1177/1609406918812346. Checked 11 Aug 2021.

Perkins, Dana L., & Swetnam, Thomas W. (1996). A dendroecological assessment of whitebark pine in the sawtooth–salmon river region, Idaho. *Canadian Journal of Forest Research, 26*(12), 2123–2133. doi: 10.1139/x26-241. Checked 11 Aug 2021.

Pfister, R. D., Kovalchik, B. L., Arno, S. F., & Presby, R. C. (1977). *Forest habitat types of montana*. Gen. Tech. Rep. INT-34. USDA Forest Service, Intermountain Forest and Range Experiment Station.

Polley, H., Wayne, Derek W., & Bailey, Robert S., & Stafford-Smith, M. (2017). Ecological consequences of climate change on rangelands. In D. D. Briske (Ed.), *Rangeland systems. Springer series on environmental management* (pp. 229–260). Cham: Springer International Publishing. doi: 10.1007/978-3-319-46709-2_7. Checked 11 Aug 2021.

Resler, L. M., & Tomback, D. F. (2008). Blister rust prevalence in krummholz whitebark pine: Implications for treeline dynamics, northern Rocky Mountains, MT, USA. *Arctic, Antarctic, and Alpine Research, 40*(1), 161–170.

Robbins, C. T., Schwartz, C. C., Gunther, K. A., & Serveen, C. (2006). Grizzly bear nutrition and ecology studies in Yellowstone National Park. *Yellowstone Science, 14*(3), 19–26.

Six, D. L., & Bracewell, R. (2015). Dendroctonus. In *Bark beetles* (pp. 305–350). Cambridge: Academic Press.

Sniezko, R. A., Kegley, A., Danchok, R., & Long, S. (2018). *Blister rust resistance in whitebark pine (Pinus Albicaulus)—Early results following artificial inoculation of seedlings from Oregon, Washington, Idaho, Montana, California, and British Columbia seed sources*. RMRS, P-76. Washington DC: USDA Forest Service.

SQCC (2021). Séliš-Qlispé Culture Committee. *Confederated Salish and Kootenai tribes.* Retrieved from http://www.csktsalish.org. Checked 11 Aug 2021.

Tomback, D. F., & Achuff, P. (2010). Blister rust and western Forest biodiversity: Ecology, values and outlook for White pines: Blister rust and western Forest biodiversity. *Forest Pathology, 40*(3–4), 186–225. doi: 10.1111/j.1439-0329.2010.00655.x. Checked 11 Aug 2021.

Tomback, D. F. (1989). The broken circle: Fire, birds and whitebark pine. In T. Walsh (Ed.), *Wilderness and wildfire* (pp. 14–17. Miscellaneous Publication 50. Missoula, MT: University of Montana, School of Forestry, Montana Forest and Range Experiment Station.

——— (2018). Whitebark pine status and the potential role of biotechnology in restoration. *Webinar.* Committee on Forest Health and Biotechnology. Retrieved from http://nas-sites.org/dels/files/2018/03/Diana-Tomback-Presentation.pdf. Checked 11 Aug 2021.

Tomback, D. F., Anderies, A. J., Carsey, K. S., Powell, M. L., & Mellmann-Brown, S. (2001). Delayed seed Germination in Whitebark pine and regeneration patterns following the Yellowstone fires. *Ecology, 82*(9), 2587–2600. doi: 10.1890/0012-9658(2001)082[2587:DSGIWP]2.0.CO;2. Checked 11 Aug 2021.

Tomback, D. F., & Kendall, K. (2001). Biodiversity Losses: A down- Ward Spiral. In D. F. Tomback, S. F. Arno & R. E. Keane (Eds.), *Whitebark pine communities: Ecology and restoration.* Washington DC: Island Press.

Tomback, D. F., & Linhart, Y. B. (1990). The evolution of bird-dispersed pines. *Evolutionary Ecology, 4*(3), 185–219.

United States Census Bureau (2012). *2010 census American Indian and Alaskan native summary file.* United States Census Bureau. Retrieved from https://www.census.gov/newsroom/releases/archives/news_conferences/20121210_aian_webinar.html. Checked 11 Aug 2021.

United States Department of Agriculture (1975). *Soil taxonomy, A basic system of soil classification for making and interpreting soil surveys.* Agriculture Handbook 436. Washington DC: United States Department of Agriculture.

USDA-NRCS (2019). PLANTS database, plant list of accepted nomenclature, taxonomy, and symbols. Retrieved from https://plants.usda.gov/home. Checked 11 Aug 2021.

United States Department of Agriculture-USFS (2021). White pine blister rust. *High Elevation White Pines.* Retrieved from https://www.fs.fed.us/rm/highelevationwhitepines/Threats/blister-rust-threat.htm. Checked 11 Aug 2021.

United States Geological Survey (2021). Predicting climate change impact on fish. Retrieved from https://www.usgs.gov/special-topic/drought/science/climate-change-impacts-fish?qt-science_center_objects=0#qt-science_center_objects. Checked 11 Aug 2021.

White, Germaine, & CSKT (2010). *Fire on the land: Native peoples and fire in the northern Rockies: Salish and Kootenai tribes fire History Project.* Lincoln, NE: University of Nebraska Press.

Whitlock, C., Cross, W., Maxwell, B., Silverman, N., & Wade, A. A. (2017). *2017 Montana climate assessment.* Bozeman & Missoula, MT: Montana State University and University of Montana, Montana Institute on Ecosystems. Retrieved from http://montanaclimate.org/chapter/title-page. Checked 11 Aug 2021.

Zhao, Chuang, Liu, Bing, Piao, Shilong, Wang, Xuhui, Lobell, David B., Huang, Yao et al. (2017). Temperature increase reduces global yields of major crops in four independent estimates. *Proceedings of the National Academy of Sciences, 114*(35), 9326–9331. doi: 10.1073/pnas.1701762114. Checked 11 Aug 2021.

8 It Is People Who Implement Climate Actions

Brenda Groskinsky

CONTENTS

8.1   LISTEN TO WHAT PEOPLE HAVE TO SAY

We all communicate, but do we understand each other? Part of the misunderstandings between groups of people is created by a number of hurdles, such as language, culture, behavior and how to receive what people are trying to say to each other.

During my career, as a Science Policy Advisor for a regional office of the U.S. EPA, I learned many vital lessons from stakeholders. One in particular is, *What People Value is What They Do.* For example, clear skies and wide-open prairies are valued significantly by Kansas natives. And they certainly do not want the prairie to succumb to the effects of climate change. So, the people who live in and around the prairies work hard to protect what they know is special.

I've also learned that being able to understand what people value makes a huge difference in how we communicate with each other. Our *Climate Action* book has authors from all over the world. One thing that we do know is that people typically value the places where they live. They are the keepers of their special places.

Finding out why they value the places where they live is critically important. It could be the most important observation that we can make during this time of climate change. Understanding what people value is imperative in the creation and implementation of specific "climate actions" in the "local" and "practical" places where they live.

After reading the chapters in this book, I hope it was very easy to witness the authors' passions for their histories, their cultures and their ideas of how to move forward in our struggles with climate change. We want to continue to treasure the

DOI: 10.1201/9781003048701-8

special places that we have grown-up in. We are all starting to realize, as well as acknowledge, that we have to save our local and practical places for our children and all of the lives that follow.

The good news is that a collection of climate actions truly has the potential to facilitate creative synergy. The more we implement climate actions, the more others will follow.

8.2   WHY IS IT SO HARD TO REVERSE CLIMATE CHANGE?

On June 23, 1988, Dr. James Hansen, the director of the National Aeronautics and Space Administration (NASA), spoke to the U.S. Congress: "Global warming has reached a level such that we can ascribe with a high degree of confidence a cause-and-effect relationship between the greenhouse effect and observed warming. ... It is already happening now" (Shabecoff, 1988).

Have you asked yourself, why is it so hard to reverse climate change? If I look back to the early 1970s, when I was a child, the words "climate change" kind of seemed like a hoax. But once I was an adult, and an employee of U.S. EPA, in the 1990s, I really started to ponder. There were just too many extreme weather events occurring and they were creating damage to homes and buildings as well as peoples' lives.

However, at that time, it was very difficult to convince my managers and environmental colleagues that the extreme weather events were clues to something more extravagant. I spent a lot of my free time trying to investigate how and why the extreme weather-related events were occurring. And then finally, and recently, I came upon a philosophical set of innovations.

> That political power must be constrained by the knowledge of climate scientists, that is, that our democracy must also be a rule of the knowers, and "epistocracy" (Plato); that such power must also be constrained by care or love of the ecological whole of which we are inevitable parts (St. Augustine); that we must not shrink from the job of enhancing the design of the "technosphere" (Descartes); that we must learn to see the whole Earth as a system and act to preserve its internal complexity (Spinoza); and that we must grant *rights* to anyone or anything—ultimately the Earth system itself— whose vital interests are threatened by the effects of climate change (Hegel).
>
> **(Williston, 2021)**

For the last ten years, I've been living in a homeowners' association (HOA). It is very apparent to me that many of my neighbors value the places where they live. However, when I requested to install solar panels on my house, as a HOA homeowner, the HOA was not supportive and neither was the local energy company. The political and financial hurdles, just to have solar panels, became somewhat overwhelming. I'm committed to persevere.

Notably, I purchased a hybrid car last year (2020). I'm really enjoying that I don't have to fill up on gas once a week. Now, I usually only fill up once a month or so. Little pieces of climate actions, that continue over time, can add up to a lot of carbon dioxide reductions.

8.3 OBSERVATIONAL COMMENTS FROM THE CHAPTER AUTHORS

As Byron Williston suggests, we should realize "the whole Earth as a system and act to preserve its internal complexity" (Williston, 2021). Perhaps we need to acknowledge and follow the three pieces of advice from the agricultural experts from Chapter 2. The authors document climate action opportunities for crop and soil management practices, for use in multiple countries. They propose an initial strategy using incentive-based opportunities with no, or minor, regulatory components. Second, they document regulatory-dominated opportunities with government exercising authority over producer practice options. Lastly, they describe implementation of long-term planning, addressing spatial and temporal land management elements, and adoption of those plans with a combination of government support and regulatory authority.

Chapter 3 comments that we, as humans, do have fears related to climate change, but more importantly, it seems that humans just really want to procrastinate. As an example, it has taken over 30 years for humans to realize that we really need to wear our seatbelts while we are driving around town in our cars! How long will it take us to implement climate actions?

The implementation of energy efficiencies, in low-income neighborhoods, is an important and necessary climate action. Unfortunately, many people procrastinate the implementation of energy efficiencies in buildings and homes, when they could be creating mechanisms to resolve the social injustices.

Chapter 4 documents how both direct and indirect climate changes cause notable impacts on grasslands. Grasslands are globally important and provide a number of ecosystem services. The scientists from Kansas State University have outlined an action plan for grasslands at the individual, community and global levels. Billions of people worldwide gain enormous ecological and social value from grasslands.

Our water systems are in trouble and solutions are becoming more and more necessary in this time of the extreme events of climate change. As an example, the Chapter 5 authors articulate the need to have the water agencies in California collaborate. There are more than 50 water supply agencies in the San Francisco area working on their own. Notably, the authors of Chapter 5 realize that stormwater capture and water reuse is going to be necessary and will require effective regional collaborative networks among all of the agencies. They note that it is important to have an equitable distribution of resources at the federal level.

The state of Minnesota has a very unique geography and requires very specific strategies for protection of their three major ecosystems (prairie, northern deciduous and boreal forests), noting they are located in several climate zones. The Minnesota Land Trust authors of Chapter 6 have documented the strategies for over 76% of Minnesota's landscape, which notably is owned by private landowners. Clearly, landowners play a critical role in the protection of their properties, especially in this time of climate extreme events.

The Confederated Salish and Kootenai Tribes (CSKT) document how they plan to save their sacred whitebark pine forest from the effects of climate change. And as they state in Chapter 7, "through the eyes of Native wisdom and western science, to conserve and restore whitebark pine and revitalize cultural relationships between humans and the biotic communities that support them" (p. 35).

LITERATURE CITED

Shabecoff, Philip (June 24, 1988) Global warming has begun, expert tells senate. *The New York Times.* https://www.nytimes.com/1988/06/24/us/global-warming-has-begun-expert-tells-senate.html Checked Aug. 28, 2021.

Williston, Byron (2021) *Philosophy and the climate crisis – how the past can save the present.* Abingdon, Oxon: Routledge.

Index

Printed in the United States
by Baker & Taylor Publisher Services